NEC3 Practical Solutions

NEC3 Practical Solutions

Robert Alan Gerrard
BSc (Hons), FRICS, FCIArb, FCInstCES

and

Stuart Kings
BSc (Hons), PhD, MRICS, MCIPS

Published by ICE Publishing, One Great George Street, Westminster, London SW1P 3AA.

Full details of ICE Publishing sales representatives and distributors can be found at: www.icebookshop.com/bookshop_contact.asp

First published 2015
Reprinted 2016, 2017, 2018, 2019, 2020

Other titles by ICE Publishing:

Managing Reality: A Practical Guide to Applying NEC3, 2nd Edition. 5-volume set.
B. Trebes and B. Mitchell. ISBN 978-0-7277-5716-6
NEC3 Engineering and Construction Contract: A User's Guide.
J. Broome. ISBN 978-0-7277-4109-7
Choosing the Right NEC Contract. T. W. Weddell. ISBN 978-0-7277-3383-2
NEC2 and NEC3 Compared. R. A. Gerrard. ISBN 978-0-7277-3384-9

www.icebookshop.com

A catalogue record for this book is available from the British Library

ISBN 978-0-7277-5969-6

© Thomas Telford Limited 2015

ICE Publishing is a division of Thomas Telford Ltd, a wholly-owned subsidiary of the Institution of Civil Engineers (ICE).

All rights, including translation, reserved. Except as permitted by the Copyright, Designs and Patents Act 1988, no part of this publication may be reproduced, stored in a retrieval system or transmitted in any form or by any means, electronic, mechanical, photocopying or otherwise, without the prior written permission of the Publisher, ICE Publishing, One Great George Street, Westminster, London SW1P 3AA.

This book is published on the understanding that the author is solely responsible for the statements made and opinions expressed in it and that its publication does not necessarily imply that such statements and/or opinions are or reflect the views or opinions of the publishers. Whilst every effort has been made to ensure that the statements made and the opinions expressed in this publication provide a safe and accurate guide, no liability or responsibility can be accepted in this respect by the author or publishers.

While every reasonable effort has been undertaken by the authorsand the publisher to acknowledge copyright on material reproduced, if there has been an oversight please contact the publisher and we will endeavour to correct this upon a reprint.

Commissioning Editor: Amber Thomas
Production Editor: Rebecca Taylor
Market Development Executive: Elizabeth Hobson

Typeset by Academic + Technical, Bristol
Printed and bound in Great Britain by Bell and Bain, Glasgow

Contents

	Preface	xii
	Abbreviations	xiii
	About the authors	xiv

01 General — 1

1.	Dealing with head office overheads in the TSC (clause 11.2(5))	1
2.	What is Site Information? (clause 11.2(16))	2
3.	What is a Subcontractor? (clause 11.2(17))	4
4.	Are in-house designers Subcontractors? (clause 11.2(17))	5
5.	How do we sign the contract? (clause 12)	6
6.	Mutual agreement to revise the Activity Schedule (clause 12.3)	7
7.	Instructing an additional *section* of the *works* (clause 12.3)	8
8.	Should we consolidate tender documentation? (clause 12.4)	9
9.	What is the status of verbal instructions? (clause 13.1)	10
10.	Can you confirm verbal instructions? (clause 13.1)	11
11.	Should we act on verbal instructions? (clause 13.1)	12
12.	Can the *Supervisor* be the *Project Manager*? (clause 14)	13
13.	Do we approve or accept? (clause 14.1)	14
14.	Could extensive *Project Manager*'s instructions render our contract void? (clause 14.3)	15
15.	What is the interaction of early warnings, the Risk Register and compensation events? (clause 16)	16
16.	What do we do about the late notification of early warnings? (clause 16)	17
17.	How do we deal with the lack of early warning notifications? (clause 16)	18
18.	How do we deal with ambiguities in the ECSC? (clause 17)	20

02 The *Contractor*'s main responsibilities — 23

19.	Refusing to do what the Works Information requires (clause 20.1)	23
20.	Who is responsible for design? (clause 21.1)	24
21.	Requesting more design information (clause 21.1)	25
22.	Proceeding with the work prior to acceptance (clause 21.2)	26
23.	Requesting the design of Equipment (clause 23.1)	27
24.	Changing key people (clause 24.1)	28
25.	Not providing services and other things as stated in the Works Information (clause 25.2)	29
26.	Missing a Key Date (clause 25.3)	30
27.	Quality control of the supply chain (clause 26.2)	31
28.	Working with Others (clause 27.1)	32

03 Time — 35

29.	Can a Completion certificate also be a payment certificate? (clause 30)	35
30.	Has Completion been achieved? (clause 30.2)	36
31.	What goes onto the first Accepted Programme? (clause 31)	37
32.	Incorporating a programme into Subcontract Works Information (clause 31)	39
33.	Programme float and time risk allowances (clause 31)	41
34.	What becomes of a tender programme? (clause 31)	42

	35.	What do we do if the programme shows a wrong date? (clause 31)	43
	36.	Problems with a programme submitted at tender stage (clause 32)	45
	37.	Lack of acceptance of programmes (clause 32)	48
	38.	Showing non-implemented compensation events on a programme (clause 32)	49
	39.	Take over and effect on liability (clause 35)	50
	40.	An *Employer*'s risk? (clause 35)	51
	41.	Damage by Others after take over (clause 35.2)	52
	42.	Where use of the *works* may not constitute take over (clause 35.2)	53
	43.	The interaction of take over, Completion and delay damages (clause 35.2)	55
04		**Testing and Defects**	**59**
	44.	Frustrated *Supervisor*! (clause 40.1)	59
	45.	What does 'unnecessary' mean? (clause 40.5)	60
	46.	Charging costs incurred for repeating tests and inspections (clause 40.6)	61
	47.	Is there a snagging list? (clause 42.2)	62
	48.	Pro forma for latent Defects (clause 42.2)	63
	49.	Constraints in the Works Information (clause 42.2)	64
	50.	Dealing with a Defect (clause 43)	65
	51.	Dealing with Defects after Completion (clause 43)	66
	52.	No operational and maintenance manuals! (clause 43)	67
	53.	Disputing the Defect (clause 43)	68
	54.	Quality control in the ECC (clause 43)	69
	55.	Consequential costs of Defects (clause 43)	70
	56.	Defects arising after the Defects Certificate is issued (clause 43.3)	71
	57.	Accepting a Defect! (clause 44)	72
	58.	Not getting access to correct Defects (clause 45)	73
	59.	Whether the *Contractor* comes back or not (clause 45)	74
05		**Payment**	**77**
	60.	Making applications for payment (clause 50.1)	77
	61.	Calculating the final account (clause 50.1)	78
	62.	Assessing the amount due (clause 50.2)	79
	63.	Retaining money due to non-compliant programmes (clause 50.3)	80
	64.	No compliant programme is ever submitted (clause 50.3)	82
	65.	Retaining monies when programme not accepted (clause 50.3)	84
	66.	Pay less notices and *delay damages* in the ECSC (clause 50.3)	85
	67.	Error in the ECSC Price List (clause 50.3)	86
	68.	Difference in payment periods between ECC and ECSC (clause 51)	87
	69.	Can the *Contractor* owe money to the *Employer*? (clause 51.2)	88
	70.	Payment in the ECSC (clause 51.2)	89
	71.	Can we add corporation tax into the Fee? (clause 52.1)	90
	72.	Passing on discounts (clause 52.1)	91

06		**Compensation events**	**93**
	73.	Omitting work under ECC Option A (clause 60.1(1))	93
	74.	Inconsistencies in the Works Information (clause 60.1(1))	94
	75.	Why is accepting a Defect not a compensation event? (clause 60.1(1))	95
	76.	Are compensation events claims? (clause 60.1)	96
	77.	Stopping the work (clause 60.1(4))	97
	78.	Dealing with Others (clause 60.1(5))	98
	79.	Late decisions (clause 60.1(6))	99
	80.	Objects of interest (clause 60.1(7))	100
	81.	Not accepting the design (clause 60.1(9))	101
	82.	Tests and inspections that the Works Information requires (clause 60.1(10))	102
	83.	Unnecessary time for tests and inspections (clause 60.1(11))	103
	84.	Interpreting physical conditions (clause 60.1(12))	104
	85.	Underground mine workings? (clause 60.1(12))	105
	86.	What are physical conditions? (clause 60.1(12))	106
	87.	Weather risk (clause 60.1(13))	107
	88.	Take over (clause 60.1(15))	108
	89.	Assumptions made by the *Project Manager* (clause 60.1(17))	109
	90.	Prevention? (clause 60.1(19))	110
	91.	Poor Site Information (clause 60.2)	111
	92.	Mistakes in a Bill of Quantities (BoQ) (clause 60.6)	112
	93.	Error in the Bill of Quantities caused by the *Contractor* (clause 60.6)	113
	94.	The difference between clauses 61.1 and 61.2 (clauses 61.1 and 61.2)	114
	95.	The time bar under clause 61.3 (clause 61.3)	115
	96.	*Project Manager's* failure to respond (clause 61.4)	116
	97.	The link between early warnings and compensation events (clause 61.5)	117
	98.	Compensation events after Completion (clause 61.7)	118
	99.	Additional work after Completion (clause 61.7)	119
	100.	Obtaining different quotes for the same compensation event (clause 62.1)	120
	101.	Updating for remaining work (clause 62.2)	121
	102.	Proposed changes to which programme? (clause 62.2)	122
	103.	Accepting quotations in a timely manner (clause 62.3)	123
	104.	Failing to respond (clause 62.3)	124
	105.	What programme should we use when assessing the impact of compensation events? (clause 63.1)	125
	106.	Using the SSCC for Subcontractor's costs (clause 63.1)	126
	107.	Forecasting the Defined Cost (clause 63.1)	128
	108.	Using the *direct fee percentage* (clause 63.1)	129
	109.	Recovering cost of additional Site staff (clause 63.1)	131
	110.	Charging for staff time in a compensation event (clause 63.1)	132
	111.	People costs in a compensation event quotation (clause 63.1)	133
	112.	Are compensation event quotations based on actual cost? (clause 63.1)	135

113.	Compensation event quotations and Subcontractors (clause 63.1)	136
114.	People not working in the Working Areas (clause 63.1)	138
115.	Extensions of time? (clause 63.3)	139
116.	Dominant cause of delay (clause 63)	140
117.	How does the contract deal with multiple programmes that have not been accepted? (clause 63)	142
118.	Including risk in compensation events (clause 63.6)	144
119.	Inconsistencies in the *Contractor's* Works Information (clause 63.8)	145
120.	Unclear Works Information (clause 63.8)	146
121.	*Contractor* not assessing compensation events (clause 64.1)	148
122.	When is a compensation event 'implemented'? (clause 65.1)	149
123.	What is an implemented compensation event? (clause 65.2)	150
124.	Having your cake and eating it! (clause 65.2)	151
125.	Can you change an accepted quotation for a compensation event? (clause 65.2)	152

07   Title   **153**

126.	Does title mean ownership? (clause 70)	153
127.	Vesting Plant and Materials (clause 70.2)	154
128.	Title to materials from excavation (clause 73.2)	155

08   Risks and insurance   **157**

129.	Weather … it's a compensation event (clause 80.1)	157
130.	A hypothetical situation (clause 80.1)	158
131.	Copper theft! (clause 80.1)	159
132.	Is terrorism a compensation event? (clause 80.1)	160
133.	Take over versus Completion (clause 80.1)	161
134.	How to define take over (clause 80.1)	162
135.	Insurance of existing buildings (clause 84.2)	163
136.	Damage caused during a search (clause 84.2)	164
137.	Accepting insurance policies (clause 85)	165
138.	Failing to provide insurance certificates (clause 85)	166

09   Termination   **169**

139.	Is there a standard ECC termination certificate? (clause 90)	169
140.	Is there a standard TSSC termination certificate? (clause 90.1)	170
141.	Can you terminate due to poor performance? (clause 91)	171
142.	Can we use Reason 21 termination due to minor earthquake? (clause 91.7)	172
143.	Can you terminate for convenience? (clause 93.2)	173

10   The Options   **175**

144.	W1 or W2? (Options W1 and W2)	175
145.	W1 or W2, which do I use? (Options W1 and W2)	176
146.	Subcontracting and inflation (Option X1)	177
147.	Indices calculations (Option X1)	178
148.	Adjusting for the price adjustment (clause X1.5)	179
149.	I think I've found a drafting error! (Option X2)	180
150.	Is this a change in the law? (Option X2)	181

151.	Applying X2 (Option X2)	182
152.	Paying in different currencies (Option X3)	183
153.	Parent company guarantee (Option X4)	184
154.	Is there a standard parent company guarantee form? (Option X4)	185
155.	Defining *sectional* Completion (Option X5)	186
156.	How is *sectional* Completion confirmed? (Option X5)	187
157.	*Sectional* Completion – effects of delay (Option X5)	188
158.	Defects date running from Completion of the whole of the *works* (Option X5)	189
159.	How to calculate the bonus (Option X6)	190
160.	Applying delay damages (Option X7)	191
161.	Dealing with late Completion (Option X7)	192
162.	Delay damages and take over (Option X7)	193
163.	Delay damages and gain share (Option X7)	194
164.	Applying partnering principles (Option X12)	195
165.	Getting performance bonds in place (Option X13)	196
166.	Paying in advance (Option X14)	197
167.	Delayed advanced payment! (Option X14)	198
168.	Design liability (Option X15)	199
169.	Reasonable skill and care (Option X15)	200
170.	Subcontractor retention (Option X16)	201
171.	Retention percentage (Option X16)	202
172.	Exceeding the retention amount (Option X16)	203
173.	Deciding the *retention-free amount* (Option X16)	204
174.	Release of retention (Option X16)	205
175.	How to draft low performance damages (Option X17)	206
176.	Limiting the liability (Option X18)	207
177.	End of liability date (Option X18)	208
178.	Providing incentive (Option X20)	209
179.	Using X20 (Option X20)	210
180.	Is a 'day' a day? (Option Y(UK)2)	211
181.	The purpose of Y(UK)3 (Option Y(UK)3)	212
182.	Preparing quotations for compensation events (clause 11.2(22))	213
183.	Increase of resources by the *Contractor* (Option C)	214
184.	Disallowed Cost (Option C)	215
185.	Productivity of resources (Option C)	216
186.	Accrued cost in an application for payment (Option C)	218
187.	The use of the Activity Schedule (Option A)	219
188.	Items on an Activity Schedule not undertaken (Option A)	220
189.	Additional management staff (Option C)	221
190.	Revising the Activity Schedule (Option A)	222
191.	Interest paid to a Subcontractor (Option C)	223
192.	What are the *Contractor*'s share payment timings? (clause 53.3)	224
193.	Purpose of Z clauses (Option Z)	225
194.	Deletion of clause 63.8 (Option Z)	226
195.	Reducing the time bar (Option Z)	227
196.	Deleting physical conditions and weather (Option Z)	228

	197.	Complying with the law (Option Z)	229
	198.	Stepping down Z clauses (Option Z)	230
	199.	Mutual trust and co-operation (Option Z)	231
	200.	Spirit of the contract (Option Z)	232
	201.	Amending a core clause (Option Z)	233
	202.	Replacing sentences (Option Z)	234
	203.	Protecting clients (Option Z)	235
	204.	Using other main Option clauses? (Option Z)	236
	205.	Valid Z clauses? (Option Z)	237
11		**Schedules of Cost Components**	**239**
	206.	Time working within the Working Areas (SCC item 1)	239
	207.	Can we recover congestion charges? (SCC item 13(a))	240
	208.	Meeting the requirements of the law (SCC item 13)	241
	209.	Using estimated staff costs (SCC item 13)	242
	210.	Payment for pension shortfall (SCC item 13)	243
	211.	Dealing with medical expenses (SCC item 13)	244
	212.	Hand tools not powered by compressed air (SCC item 44)	245
	213.	Do we get the Working Areas overheads on top of Subcontractors' costs? (SCC item 44)	246
	214.	Is security in the Working Areas overheads percentage? (SCC item 44)	247
	215.	How is a security guard paid for? (SCC item 44)	248
	216.	Working Areas overheads percentage on people (SCC item 44)	249
	217.	What does 'services' mean in the SCC? (SCC item 44)	250
	218.	Using the SCC Working Areas overheads percentage (SCC item 44)	251
	219.	Damage to Equipment (SCC item 7)	252
	220.	Damage to vehicles (SCC item 7)	253
	221.	Can we recover the costs of transport for people? (SSCC item 1)	254
	222.	Cost of people in the SSCC (SSCC item 1)	255
	223.	Which SSCC Equipment rates do we use? (SSCC item 2)	256
	224.	Cost of Equipment standing (SSCC item 2)	257
	225.	Using the percentage for people overheads (SSCC item 41)	258
	226.	Using the SSCC percentage for people overheads (SSCC item 41)	259
	227.	Using the percentage for design overheads (SSCC item 6)	260
12		**Contract Data**	**263**
	228.	What *services* are in the PSC Scope? (Contract Data)	263
	229.	Who is Mr I Judge? (Contract Data)	264
	230.	How do you describe the Affected Property? (Contract Data)	265
	231.	Incomplete Contract Data (Contract Data)	266
	232.	Do we need to name the *Project Manager*? (Contract Data)	267
	233.	Works Information versus the *works* (Contract Data)	268
	234.	Best practice in drafting the Contract Data (Contract Data)	269
	235.	The *period for reply* (Contract Data)	270
	236.	Responding quicker than the *period for reply* (Contract Data)	271
	237.	Contractual significance of the Risk Register (Contract Data)	272

238.	The *defects date* versus the *defect correction period* (Contract Data)	273
239.	Does the *defects date* change? (Contract Data)	274
240.	Optional statements – what are they? (Contract Data)	275
241.	Defining key people (Contract Data)	276
242.	Modelling the Contract Data part two rates and percentages (Contract Data)	277
243.	Optional statements (Contract Data)	278
244.	Completing the Contract Data (Contract Data)	279
245.	Changes to the Contract Data part two percentages (Contract Data)	280

Preface

It is almost a decade since the third version of the NEC suite of contracts was published. As well written as the NEC3 contracts are, there is a real appetite from users to obtain a better understanding of these contracts than others before, and therefore numerous questions as to their practical application prevail. When analysed, it is evident that similar questions tend to be repeated – highlighting a potential gap in users' knowledge that might benefit from better guidance or more training.

NEC3 Practical Solutions provides a comprehensive set of responses to common questions posed on the NEC3 suite of contracts. It is not therefore a clause-by-clause analysis, but addresses the more common issues, problems and queries that arise in use.

We have based this book on real questions posed to the NEC Users' Group helpline over the past few years. We noticed that there were certain questions that were arising quite frequently, particularly concerning defined terms, the programme, payment, compensation events and Defined Cost. It is these recurring questions that have formed the basis for the book. *NEC3 Practical Solutions* provides detailed contractual responses to almost 250 common queries faced by practictioners during the design and implementation stages of a project, as well as the risks associated with running a project. As it mirrors the typical questions posed, there is a natural focus on the NEC3 Engineering and Construction Contract (ECC). However, the book also covers other contracts within the NEC3 suite.

The book is structured in the same chronological order as the ECC, which we hope will enable readers to use the book as a quick look-up tool to provide answers to common questions and problems that occur in practice. We have also included tips on better drafting of contract documents and made reference to further guidance to enhance its usefulness.

We wanted to write the book because, as NEC3 Consultants, we recognise that the same repeat problems tend to emerge on projects. We are both passionate about providing concise and practical solutions to these common problems in a format that can be readily sourced. In turn, this should help project teams resolve differences in contract interpretation quickly.

It is our hope that *NEC3 Practical Solutions* will be an invaluable reference guide for engineers, surveyors, architects, project managers, supervisors, consultants, contractors, subcontractors and anyone else working with or interested in working successfully with the NEC3 suite of contracts.

Abbreviations

AC	NEC3 Adjudicator's Contract (April 2013)
ECC	NEC3 Engineering and Construction Contract (April 2013)
ECS	NEC3 Engineering and Construction Subcontract (April 2013)
ECSC	NEC3 Engineering and Construction Short Contract (April 2013)
ECSS	NEC3 Engineering and Construction Short Subcontract (April 2013)
FC	NEC3 Framework Contract (April 2013)
PSC	NEC3 Professional Services Contract (April 2013)
PSSC	NEC3 Professional Services Short Contract (April 2013)
SC	NEC3 Supply Contract (April 2013)
SCC	Schedule of Cost Components
SSC	NEC3 Supply Short Contract (April 2013)
SSCC	Shorter Schedule of Cost Components
TSC	NEC3 Term Service Contract (April 2013)
TSSC	NEC3 Term Service Short Contract (April 2013)

About the authors

Robert Alan Gerrard
BSc (Hons), FRICS, FCIArb, FCInstCES
Robert is a chartered quantity surveyor with 30 years' experience of setting up and managing contracts, mainly in the construction industry, in the UK and overseas.

He has been the NEC Users' Group Secretary since 2005 and has authored some of the documents in the NEC3 family.

Robert is a full-time NEC consultant and has run numerous NEC training courses, working with NEC contracts since 1996.

Dr Stuart Kings
BSc (Hons), PhD, MRICS, MCIPS
Stuart provides training and consultancy throughout the world on NEC3 and has authored numerous articles. He gained his doctorate in improving the effectiveness of tender documentation.

Through work in all industry sectors he has gained a detailed practical knowledge of the NEC3 suite of contracts and the ingredients to successful project delivery.

He has recently developed the NEC3 ECC Accredited Project Manager programme for the NEC.

Stuart is also a director and co-founder of Sypro Management Ltd. (www.sypro.co.uk), which specialises in on-line NEC3 contract management.

Chapter 1
General

1. Dealing with head office overheads in the TSC (clause 11.2(5))

Q How do we deal with head office overheads when using the TSC, are these covered in the Fee?

A There is no direct reference in the TSC to 'head office overheads' and this term might mean different things to different people. You need to understand what cost the *Contractor* can recover through the contract – this is called Defined Cost and is defined in clause 11.2(5). Clause 52.1 then confirms that all costs not included in Defined Cost are assumed to be in the Fee.

So you need to concentrate on the four headings in the Defined Cost definition and understand what they mean/cover – they are people, Plant and Materials, work sub-contracted and Equipment. It does not matter therefore whether, for example, a buyer is in the head office, or in the local office, or in the Affected Property as such, so there is no limit as to where people are undertaking the *services*, as long as they are working on the *services*. Such cost would be a part of Defined Cost. As you work through all other costs on the same basis, you will appreciate what cost falls within the definition of Defined Cost and that everything else is deemed to be included in the Fee.

2. What is Site Information? (clause 11.2(16))

I have a query regarding the content of Site Information in an ECC Option A contract used for design-and-build procurement. The project involves the design and construction of an asset, but the Site is to be determined by a developer and will be surrounded by a housing development and highways, which are yet to be designed in detail. Does the source of information impact its status as Site Information? For example, if a ground investigation or similar report is procured by the design-and-build *Contractor*, does it still constitute as being Site Information? Similarly, if a ground investigation report is provided by a third-party housing developer, could that be considered to be Site Information?

The first thing to say is that Site Information is a defined term (see clause 11.2(16)), so only information that falls within that definition will count. Part of the definition refers to 'documents which the Contract Data states it is in', which means it has to be referenced in the relevant entry in Contract Data part one.

So, to answer your question, the source of the information in the Site Information is irrelevant. If it is in the Site Information in the contract and complies with the first bullet of clause 11.2(16), then it is Site Information. The Site Information is, however, used in the ECC for only one purpose, which is to help determine what 'an experienced contractor would have judged at the Contract Date to have such a small chance of occurring that it would have been unreasonable for him to have allowed for them' – see clauses 60.1(12), 60.2 and 60.3. Site Information is important in that it enables the *Contractor* to make reasonable assumptions at tender stage, rather than assuming the worst and pricing for it.

You refer to a ground investigation procured by the *Contractor*. The status of that will depend on when it is procured and what the contract says. If the ground investigation is procured at tender stage, and if the *Contractor* bases its price on the investigation, then it is sensible to include the investigation within the Site Information in the subsequent contract entered into. That gives better certainty to both Parties. In those circumstances, however, the *Employer* should satisfy itself that the ground investigation has been competently carried out as it potentially takes the risk for any physical conditions that are worse than shown. Of course the benefit to the *Employer* is that the tender price should be lower as a result.

There is no facility in the contract to change the Site Information, unlike the Works Information. Once the contract is let, any ground investigation sourced by the *Contractor* does not become Site Information. Of course the *Contractor* is entitled to argue that the information in its later ground investigation shows that the physical conditions are such as to trigger a compensation event under clause 60.1(12), but that will be judged partly on any Site Information you put into the contract.

Finally, NEC has produced a guide on how to write the Works Information for the ECC, which also includes a section on the difference between Site Information and Works Information. We strongly recommend that anybody tasked with putting together both Works Information and Site Information read this document.

3. What is a Subcontractor? (clause 11.2(17))

Q We are using the ECC and in relation to clause 11.2(17) are trying to understand the cut-off between labour-only operatives and Subcontractors. Where is the line drawn?

A What you have to look at is the contract between the *Contractor* and Subcontractor. If it falls within one of the bullets stated in clause 11.2(17) then they are a Subcontractor, otherwise they are not. So, if they have a contract to lay all the brickwork on a labour-only basis, for example, and are paid an amount per square metre to do so, then they are Subcontractors under the first bullet. In that case they have the obligation to provide the resources to do it within the time set out in the contract. If they instead have a contract to supply bricklayers at an amount per hour (and equipment at an amount per hour), and those bricklayers will be put to work as determined by the *Contractor*, then they are not Subcontractors because their obligation is to provide labour to work rather than the end result.

General

4. Are in-house designers Subcontractors? (clause 11.2(17))

We have been on a fast track ECC Option E contract and the design was developed as the works proceeded. As a result the designer had a heavy presence on Site as well as backup in the design office. The rates within Contract Data part two are now 5 years old and the contract has an inflation provision for them. The designer is 'in house', albeit he or she is a separate legal entity under the umbrella of the parent company. There was no formal subcontract in place.

Within the Contract Data part two there is a completed table for design employees (disciplines) under the heading 'The hourly rates for Defined Cost of design outside the Working Areas are'. The percentage for design overheads is 17.5%. The table that is entitled 'The categories of design employees whose travelling expenses to and from the Working Areas are included in Defined Cost' has been left blank.

Under the ECC, should the designer be treated as a Subcontractor? If not, should the design works (done outside the Working Areas) be valued at the hourly rates in Contract Data part two, with design overhead + design overhead + fee + inflation? Also, if not a Subcontractor, should the design works done inside the Working Areas be valued at the same hourly rates + travelling expenses + design overhead + fee + inflation?

These designers are not a Subcontractor because you have not got a subcontract with them – see clause 11.2(17). It is difficult to comment on the inflation provision as no further details are provided. For design work carried out outside the Working Areas you will be paid for the designer's rate plus the percentage for design overheads in the Contract Data part two plus the Fee. You will not get travelling expenses because you have not listed anybody who is entitled to them (see item 63 in the SCC).

The designers' rates only apply for design carried out outside the Working Areas (see item 6 of the SCC). If they are carrying out design in the Working Areas your quoted rates are not used at all. Instead you use the calculation for people set out in items 11, 12 and 13 of the SCC and to that you add your Working Areas overheads percentage in Contract Data part two (see item 44 of the SCC) plus, as always, the Fee.

5. How do we sign the contract? (clause 12)

We have a copy of the ECC, but there is nowhere for the Parties to sign. Is there another document that sits with this, or do we have to create our own signing provisions?

The omission is quite deliberate. It is for the Parties to decide how the contract is to be made, because it could well depend on the law of the contract (or even the law of the country one or more of the Parties reside in). In the UK, it could also depend on whether you want a deed, or not.

So, you can make it as simple or as complicated as the applicable law requires. Simple offer and acceptance will do in the UK, or you can write out a more formal document (deed or not) for both Parties to sign.

6. Mutual agreement to revise the Activity Schedule (clause 12.3)

We have an ECC Option A project on Site where the programme has been revised. As funding is time critical, we are questioning how we can ease the release of payments. Considering clause 32.1, can the *Project Manager* and *Contractor* mutually agree to revise the Activity Schedule to co-ordinate with the accepted revised programme including splitting original contract activities within the Activity Schedule and tasks within the revised programme into two sub-activities (for example, 'Delivery of materials on Site' and 'final complete installation')?

The only way that you can change the Activity Schedule is as a result of a compensation event (clause 63.12) or if the *Contractor* changes its planned method of working (clause 54.2). Neither applies in this case. The only way this can be achieved is if the *Employer* and *Contractor* agree to change the contract and it is confirmed in the way set out in clause 12.3.

7. Instructing an additional *section* of the *works* (clause 12.3)

I am the *Project Manager* working on an ECC design-and-build contract. The *Employer* has instructed me to instruct the *Contractor* to accelerate a certain amount of work to ensure it is completed on a certain date, which is before the current Completion Date.

Is the right way to do this to notify the *Contractor* of a new *sectional* Completion Date, providing the *Contractor* a list of the scope of work to be completed by this date and then ask the *Contractor* for a quotation for the impact of this?

You simply do not have the power under the contract to issue an instruction to complete works by a certain date or add a new *sectional* Completion Date to an existing contract. If you tried to do that the Contractor could ignore your instruction with impunity, because they are only required to obey an instruction that is given in accordance with the contract (clause 27.3).

All that can happen is that the *Employer* (not the *Project Manager*) and *Contractor* can agree to change the contract to accommodate the new requirements. It is likely that the *Contractor* will only agree to this for an additional sum, payment of which will form part of the agreement. Any such agreement must comply with clause 12.3. We therefore recommend that the *Employer* needs to start talking to the *Contractor* about this, because the longer the agreement takes the more expensive it is likely to be. As this only applies to a certain amount of the *works* rather than all of it you cannot even use clause 36. And even if you could, you still need to get the *Contractor*'s agreement to the quotation before proceeding.

8. Should we consolidate tender documentation? (clause 12.4)

We are about to start a contract using ECC Option B, which has been the subject of a prolonged period (6 to 10 months) of exchanges, refinements and discussions following the initial tender. We think it would be beneficial for a copy of a definitive document, which presents all the agreement after all the correspondence and negotiations, to be supplied before commencement. We consider this would make the contract management task simpler for both ourselves and the *Project Manager*. We have asked for this but have been advised along the lines that we have all the information necessary already and additionally that it will be available by means of the electronic management system used for tendering. It would appear that the intention is to work with the multiple volumes of all exchanges/offers/counter-offers through the period as a package of emails rather than presenting the out-turn in a more concise package that references the necessary changes. Is there any written requirement in ECC for a consolidation of such documentation?

We would recommend that you do not do this. For starters, there is simply nowhere for these records of your negotiations to be incorporated into an NEC3 contract. That is quite deliberate. And if they are not in the contract, then they probably will not be part of the contract (although that latter point is a legal matter that we cannot advise on). You need to start with a clear, unambiguous and consistent contract (as with all contracts) and you will not do it in the way suggested because the documents that form the negotiations will, inevitably, be inconsistent and often ambiguous, as Parties take certain lines and then subsequently concede them. Our recommendation is that once you have reached your agreement, you change the original documents (Works Information, Activity Schedule, etc.) to reflect that agreement and use those as a basis for your contract. Case law (and for that matter adjudications and arbitrations) are littered with examples of cases where people have done as your client has suggested and the Parties have then, much later, fallen out over what they think the 'agreement' they made meant. This advice applies to all contracts under all forms of contract.

9. What is the status of verbal instructions? (clause 13.1)

Q We recently finished projects using the ECC in which we have completed lots of additional work, but have not had instructions from the *Project Manager* (as verbally promised). The *Employer* is refusing to pay any further monies until we show instruction and allocation sheets. What should we do?

A Verbal instructions have no status in this contract, they have to be in a form that can be read, copied and recorded (see clause 13.1). You should not have acted on these instructions until they were in that form. And if the *Project Manager* has now proved to be dishonest by promising them in that form and then not delivering, you are potentially in trouble. As an aside, allocation sheets are not required for compensation events – you have to make an assessment as to what the changes will cost, not keep records and use them afterwards (see clause 63.1). We consider you will need to take legal advice on furthering this matter.

10. Can you confirm verbal instructions? (clause 13.1)

We are using ECC Option B and have a query on confirmation of verbal instructions. We recall that confirmation of verbal instructions may be limited and has little or no contractual value and that every instruction has to be given in writing by the *Project Manager* (or the *Supervisor* for those limited areas he or she is responsible for).

1 What is the status of confirmation of verbal instructions and what are their limitations of contractual usefulness?
2 Clause 13.1 states that communications have to be in writing so if a confirmation of verbal instruction was to be issued, and bearing in mind it is in a form that could be read, copied and recorded, would it be deemed as a contractually legitimate instruction if the *Project Manager* countersigned either simply with his or her name, or even stated 'Accepted' or 'Agreed'?

1 All communications have to be in a form that can be read, copied and recorded (clause 13.1). In addition, notifications (of early warnings, compensation events, etc.) have to be communicated separately from any other communication (clause 13.7). All of that means that there is no such thing as a valid verbal instruction and therefore no such thing as a confirmation of verbal instruction.
2 The *Project Manager* and the *Contractor* need to understand that NEC3 contracts require you to communicate professionally. The *Project Manager* should not give an instruction unless he or she is willing to put it in a form that can be read, copied and recorded, and you should not accept or act on one unless it is in that format. If the *Project Manager* is not willing to do so, then really he or she should resign immediately, because he is not following the requirements of the *Employer*'s contract. That is the only safe advice we are able and willing to give you.

11. Should we act on verbal instructions? (clause 13.1)

We have an ECC construct-only project in which the *Contractor* built a car park. The design (included in the Works Information) was provided by the *Employer*. It turns out there were problems with the design and the *works* could not be built. Although written early warning notifications were raised, there were no instructions at the time from the *Project Manager* to change the Works Information or any record of what was discussed at the risk reduction meetings. However, the *Contractor* acted on verbal instructions and revised drawings from the *Project Manager*, which changed the Works Information and the *works* were built.

A year later a compensation event was notified by the *Project Manager* as well as formally changing the Works Information to match the solution built. Therefore, there is no dispute about it being the change to the Works Information being a compensation event, just a difference in how to prepare the quotation. In relation to the quotation, when should it be deemed assessed from (i.e. what part is on actual Defined Cost and what part is on forecast Defined Cost)?

Neither party seems to have covered themselves in glory on this contract! Verbal instructions are not valid in the ECC. The *Project Manager* should not have given them and your staff should not have acted on them. It seems that the *Project Manager* issued drawings etc. for the new design and that would probably be enough to say it is correctly issued at that time. When the verbal instruction was issued the *Project Manager* should have issued the instruction correctly, should have notified this as a compensation event and instructed you to provide a quotation. This conclusion is reached reading clauses 14.3, 13.1 and 61.1 in that order.

Therefore, the date of the verbal instruction and revised drawings issued is the date you use to split 'actual' from 'forecast' Defined Cost, and to decide which Accepted Programme should be used. That means all of it is assessed using forecast Defined Cost. Quite deliberately, the switch date does not depend on when the assessment is made, because that will be different depending on who is doing the assessing, you, the *Project Manager* or the *Adjudicator*.

12. Can the *Supervisor* be the *Project Manager*? (clause 14)

We are using the ECC Option B. In this, the contract states that the *Project Manager* and the *Supervisor* should be distinct and separate; further, Managing Reality states that there should be no line reporting from the *Supervisor* to the *Project Manager* and that the *Supervisor* carries out his specified duties independently of the *Project Manager*. It further states that even if there is line reporting between the two people named in their employment conditions, outwith their ECC designated roles, this relationship cannot be mirrored in their duties carried out under the contract.

Bearing this in mind and acknowledging that the contract permits the *Project Manager* to delegate those duties he or she wishes to others, can the *Project Manager* delegate any of his or her duties to the *Supervisor*, or is there a conflict here with the intentions of how the ECC should be run, which would mean that these delegated responsibilities cannot be fulfilled by the *Supervisor*?

The roles and responsibilities are different. The skill sets are different. The *Project Manager* cannot just do in the ECC what the *Supervisor* does and vice versa. The *Supervisor* is not subservient to the *Project Manager* and the *Project Manager* cannot 'overrule' him or her when it comes to the *Supervisor*'s duties.

That said, there is no reason at all why the same person should not carry out the two roles on small/medium-sized projects. It is not at all unusual in those circumstances for the same person to be named as both *Project Manager* and *Supervisor*. Equally, the *Project Manager* can delegate some of his or her actions to the *Supervisor* should he or she wish. There is no conflict with the intentions of those drafting the contract in either of these options. Having said that, if the *Project Manager* delegates all of his or her actions to somebody else then one questions whether that other person should have been named as the *Project Manager* in the first place!

13. Do we approve or accept? (clause 14.1)

Does it matter if we continue to use the term 'approved' when we sign off things like *Contractor*'s drawings? This fits within our company QA requirements.

It is important that you follow the requirements of the particular contract you are working with. In ECC, for example, if there is design to be undertaken by the *Contractor* and submitted for acceptance by the *Project Manager*, then you are to 'accept' this, not 'approve' (see clause 21.2). If you use words in your standard processes that are not associated with the particular contract, then have you actually discharged your obligation in that contract? This is further reinforced through clause 14.1, which states such acceptance does not change the *Contractor*'s liability for his design. You may need to change your company QA requirements; you may need to discuss those requirements of the contract with your insurers that may positively impact on your premiums; you do though need to understand and follow the prescriptive requirements of each contract properly.

14. Could extensive *Project Manager*'s instructions render our contract void? (clause 14.3)

Q We have an ECC contract with such a significant change that we believe the contract to be essentially rendered void – what do you think?

A Although you give very brief details here, we cannot offer legal advice so what we can say on this matter is in any case limited. The *Project Manager* has an apparently unfettered right to change the Works Information – see clause 14.3. However, the *Contractor*'s obligation is to carry out the *works*, as defined in the Contract Data part one (see the second bullet) – see clauses 20.1 and 11.2(13).

For example, if the *works* are described as to build a power station, it could be argued that the *Project Manager* could not instruct the *Contractor* to build an entirely and unconnected water treatment works; although he or she could instruct the *Contractor* to build one to treat the water coming into or out of that power station. In these cases the *Contractor* is protected because the instruction will be a compensation event that is valued based on a cost and time forecast. In addition, the Site is defined in clause 11.2(15) and the relevant entry in the Contract Data, and neither Party has the unilateral ability to change that definition. So it could be argued that the *Project Manager* could not instruct anything to be built outside of that area.

There seems to be no doubt that the *Project Manager* can instruct extensive changes to the Works Information, but how far he or she can go will end up being a legal question, which we cannot answer. And if he or she goes too far, whether that would render the whole contract void is also a legal question involving all sorts of complex legal principles.

15. What is the interaction of early warnings, the Risk Register and compensation events? (clause 16)

We would like to get clarity on the use and updating of the Risk Register within NEC3 contracts. We have an ECC contract using Option A and the *Contractor* is claiming what we think is not an additional cost for the specified risk/compensation event.

Our contract stated that the *Contractor* can use some old apparatus for a new installation. This risk was recorded in Contract Data part one as a matter to be added to the Risk Register. Unfortunately the old apparatus was damaged by fire and cannot be re-used. A quotation was requested and the *Contractor* is claiming things like Subcontractor transportation for his people, to and from his Site and Working Areas. Our question is, how did the *Contractor* intend to do this work originally and what would be the extra in this case?

The first thing to point out is that the Risk Register and early warnings have got nothing directly to do with compensation events. They are about managing the risk of problems that may occur, whoever's they are. Compensation events are about events that have occurred or are going to occur and that are at the *Employer*'s risk. Not all early warnings will be about future compensation events, and not all compensation events will have an early warning.

We assume there was a *Project Manager*'s instruction that changed the Works Information, resulting in a compensation event arising under clause 60.1(1). The value of this compensation event is based on the forecast effect it is going to have on Defined Cost – see clause 63.1. Therefore you have to decide how much the Defined Cost has risen. If this will mean installing new apparatus (you do not say) then the additional cost (if any) of the additional people doing that will have to include any additional costs of getting them to Site. That will be part of the 'amount paid by the *Contractor*...' – see item 11 of the SSCC. It is difficult to be any more specific because we do not know what the transport arrangements are likely to be. However, if the Defined Cost of those travel arrangements increases because of this compensation event then the *Contractor* is entitled to recover that additional cost.

16. What do we do about the late notification of early warnings? (clause 16)

We have a *Contractor* on an ECC contract who is notifying early warnings well after the event. We believe they are doing this in a cynical attempt to overload the *Employer*'s management team and achieve a tactical advantage. They probably believe this will end up in adjudication. We will be able to demonstrate they have not attempted to minimise our costs at all over the duration of this contract. Can we as *Project Manager* reject an early warning and what benefit is there for any party to notify early warnings after the event?

Clause 16.1 is clear and it appears to be misunderstood here. You only give an early warning of a matter that 'could' have the listed effect. If it has already had the effect then it is not an early warning; instead it is a late warning! So the timing depends on when the effect occurred, not when the event did. You must remember that the notifying of early warnings is not a one way street. The onus is as much on you the *Project Manager* to do so if you are aware of any matter too, and if you do not you will be in breach.

Early warnings are a form of risk management. They are not compensation events and do not entitle the *Contractor* to any more time or money. Notifying an early warning is not the same as notifying a compensation event. The *Contractor* still has to do the latter, even if he has already done the former; and he has to do it within the 8 weeks set out in clause 61.3. The only proviso to that is that none of that applies to compensation events that result from your instructions, when the contractual onus is on you to notify (clause 61.1) and the 8-week time limit does not apply.

Having said all that, our view is that it is always in the *Employer*'s interests if the *Project Manager* is seen to be actively considering and discussing all early warnings whether they are strictly contractually valid or not.

17. How do we deal with the lack of early warning notifications? (clause 16)

We are using the ECS Option A. There is a provision within the contract that enables the *Contractor* to consider early warning notices (or more particularly, the lack of) when assessing compensation events using clauses 61.5 and 63.5. Assume a Subcontractor risk event occurs and an early warning is notified as it could affect the programme – clause 16.3 requires the Parties to co-operate to ensure, as far as possible, that actions are taken and decisions are made that avoid or mitigate the effects of identified risks on cost, quality and time. In this case the Subcontractor requires assistance from the *Contractor* in the induction of operatives and acceptance of method statement changes. The *Contractor* unfortunately is not forthcoming with assistance and instead waits for the Subcontractor to sort the issue, but the effects of the event could have been reduced if the assistance from the *Contractor* had been received.

With reference to the above timeline and the provision in clause 63.5, is it conceivable that the delays are extra over to those that would have been incurred anyway as a result of the risk event occurring? For example, risk event occurs, *Contractor* assistance received, total delay = 1 day. Or risk event occurs, *Contractor* assistance not received, total delay = 3 days.

In the above example, is it possible to pass the 2 days additional delay back to the *Contractor* – had the *Contractor* acted as stated in the contract this would have mitigated the delay? I am aware that there is no express clause in the contract for this, it is just the contract will punish the Subcontractor for not notifying an early warning, but does not punish the *Contractor* for not providing as much assistance as is reasonably practicable.

The main way that the Completion Date or Prices change is if a compensation event occurs. And we cannot see how one has occurred. And there is not one that says 'the *Contractor* has not helped me sort my problem out'. We assume that you were aware of the need for these inductions and acceptances – either they were in the Subcontract Works Information, or they were part of the law. In that case you were required to sort them out and they were your obligation. Please remember in all this that the early warning and compensation processes are entirely different. The clauses (61.5 and 63.5) you refer to are about you not notifying an early warning of something that was at the *Contractor*'s risk, in the sense that it would turn into a compensation event. This is something that is entirely at your risk and which you knew about, but did not manage properly. We know that sounds harsh but, as we say, we assume that you knew about it and yet you still let it delay the work, in the sense that you did not allow enough time for it. So, for the reason in our first paragraph, we cannot see how this can be a compensation event.

18. How do we deal with ambiguities in the ECSC? (clause 17)

Clause 17.1 of the ECC deals with ambiguities and inconsistencies. Can you please advise how ambiguities are dealt with under the ECSC?

The ECSC is designed for simple low-risk work, and therefore there would hopefully be fewer ambiguities. The only place this is likely to be would be in the Works Information. In that case, if it is unclear or uncertain, you should ask the *Employer* to clarify what is meant. Any instruction given will be a change to the Works Information that would be a compensation event (clause 60.1(1)). That would be a compensation event and, in assessing that compensation event, it will be assumed that you used in your tender the interpretation most favourable to you (clause 63.8) – that is the one that cost the least.

To give a practical example, if one part of the Works Information says you use a particular item that costs £100 and another specifies, for the same thing, a different item that costs £200, you will be assumed to have allowed for the cheaper one. If the *Employer* says 'I want the expensive one' then the value of the compensation will be £100 plus the stated percentage uplift (assuming all other Defined Cost is the same).

Some further thoughts and top tips

- Make sure when drafting the ECC Works Information (and the equivalent in other NEC3 contracts) that you use the same terminology as laid down in the contract (e.g. 'Defined Cost' and 'Accepted Programme').
- The *Employer* needs to think about appointing a suitable *Project Manager* and *Supervisor* with the right skills.
- Take care not to give or act on verbal instructions; follow the contract, put communications in a form that can be read, copied and recorded.
- Take care to use words such as 'acceptance' in properly managing the contract where the contract requires this.
- Appreciate that early warnings are a good and positive thing that can help in best practice contract management. They are entirely different from compensation events and cover very different things.
- Note that only the *Contractor* and the *Employer* can change a term of the contract that exists, the *Project Manager* cannot waive, amend or delete a project of another's contract.

Chapter 2
The *Contractor*'s main responsibilities

19. Refusing to do what the Works Information requires (clause 20.1)

Q The Works Information requires our *Contractor* to prepare a breakdown of cost in a particular format but they are refusing to do so. Do they have to do this? What provisions exist within the contract to enforce this?

A Clause 20.1 is a very short yet key clause – the *Contractor* Provides the Works in accordance with the Works Information. This is defined in clause 11.2(18) as information that either specifies and describes the *works* or states any constraints on how the *Contractor* Provides the Works. This is either in the documents that the Contract Data states it is in or any subsequent instruction. Clause 14.3 allows the *Project Manager* to change the Works Information. The *Contractor* is therefore obliged to comply with the Works Information and any subsequent instruction (see also clause 27.3). Ultimately, continued default of the *Contractor*'s obligations could result in termination (clause 91.2, R11). The suggested approach would be to meet (perhaps by means of an early warning), explain the provisions within the contract and encourage compliance in order to avoid the ultimate sanction within the contract.

20. Who is responsible for design? (clause 21.1)

Q We are delivering a project under ECC Option A and the project is largely designed by the *Employer*. The Works Information stated that, as *Contractor*, we should design the piling but was silent on the rest. The steelwork has been designed and calculated by the *Employer*'s structural engineer but we are debating who should be responsible for the design of the fixing details.

A The answer is simple. Under clause 21.1 the *Contractor* is only required to design the parts of the *works* that the Works Information states he is to design. In this case that is solely the piling. If the *Project Manager* now wishes you to design the fixing details then a change to the Works Information should be instructed and this will then be a compensation event under clause 60.1(1).

21. Requesting more design information (clause 21.1)

We are using Option C of the ECC and are required to design and build a new sports complex for our client. The Works Information clearly states that the *Contractor* is responsible for all of the design – which is not in contention. However, the *Supervisor* is insisting on seeing more and more of the drawings and design. This is incurring more cost for us as they are insisting on paper copies of each drawing – where do we stand with this?

One important initial comment is that the *Supervisor* has no power in the contract to request particulars of design. Clause 21.2 is the important clause in this situation. You are only required to submit particulars of the design as the Works Information requires – if it says nothing then that is exactly what you provide. If the *Project Manager* (not *Supervisor*) wishes to change the Works Information under clause 14.3, and then requests particulars of the design then you will be required to comply with this (clause 27.3). However, this change to the Works Information will be a compensation event (clause 60.1(1)). The last sentence in clause 21.2 is very important and must be complied with – it states that you do not proceed with the work until the *Project Manager* has accepted. No timescale is mentioned in this clause so we need to revert to the *period for reply* (clause 13.3). This needs to be built into your programme and assessed as part of the compensation event.

22. Proceeding with the work prior to acceptance (clause 21.2)

It is the first time we have used the ECC contract and we have opted for Option A. We followed the Works Information guidance published by the NEC in April 2013 and specified the parts of the *works* that the *Contractor* is required do design. The project is fast track and has £50 000 per day damages. As a result of this the *Contractor* is progressing with the design and work very quickly. Our concern is that the *Contractor* is undertaking work prior to our acceptance.

Our question relates to part of the design that has been submitted for acceptance. This has triggered thoughts in the *Employer* team and, in hindsight, the original Works Information needs changing. What do we do in this situation? Is this a compensation event?

What the contract says is that, assuming you requested the *Contractor* to submit particulars of their design (under clause 21.2), then they are not to proceed until the *Project Manager* has accepted. Failure to comply would be a default; continued default may result in termination of their obligation to Provide the Works (clause 91.2, R11).

This period of consideration is deliberately built into the contract to allow the *Project Manager* to review the design and, as in the latter case, potentially change your mind. The response period will be the *period for reply* (see clause 13.3). The two reasons for not accepting (note that we do not use the term 'reject') are that it does not comply with the Works Information or the applicable law. So, if you do not accept because you have changed your mind then this would in turn be a compensation event (clause 60.1(9)). It is important to meet with the *Contractor* and explain these provisions within the contract.

23. Requesting the design of Equipment (clause 23.1)

I am the *Project Manager* on a large petrochemical project and my internal health and safety manager is keen to see the design of earthwork support for deep trench excavations and also a temporary footbridge required for the project. I cannot find a clause in the contract that allows me to request this.

The clause in this case is clause 23.1. Equipment under the ECC includes items provided by the *Contractor* and used by them to Provide the Works and which the Works Information does not require them to include in the *works* (clause 11.2(7)). You can instruct (under clause 23.1) that the *Contractor* submit particulars of the design of an item of Equipment for acceptance. You can only 'not accept' if you consider that it will not allow the *Contractor* to Provide the Works in accordance with the Works Information, the *Contractor*'s design, which the *Project Manager* has accepted, or the applicable law.

24. Changing key people (clause 24.1)

We interviewed a *Contractor* as part of the tender process for an ECC project, the *Site Agent* proposed (listed as a key person) was highly experienced in the healthcare sector (which we are in). We have now heard that he has obtained a new job and is about to leave. What provisions are there in the contract for dealing with this?

Assuming that he was listed as a key person in Contract Data part two then his responsibilities, qualifications and experience should also have been detailed. Clause 24.1 requires that the *Contractor* submits the name, relevant qualifications and experience of any proposed replacement person for acceptance by the *Project Manager* – a reason for not accepting is that their relevant qualifications and experience are not as good as those of the person who is to be replaced.

25. Not providing services and other things as stated in the Works Information (clause 25.2)

We have a large bypass project under Option B of the ECC. We specifically requested that the *Contractor* provided us with a cabin on Site, internet and electrical connection. They have failed to provide this and, as a result, we are incurring additional costs. What provisions are there in the contract for dealing with this?

Clause 25.2 requires that the *Contractor* provides services and other things as stated in the Works Information. The same clause also states that any cost incurred as a result of the *Contractor* not providing is assessed by the *Project Manager* and included as a negative adjustment to the amount due under clause 50.2 ('less amounts to be paid by or retained from the *Contractor*').

26. Missing a Key Date (clause 25.3)

We have a Key Date on our ECC Option C healthcare project. This is for the construction of a concrete plinth and first fix electrics ready to take a cancer scanner from a specialist supplier with whom we have a direct contract.

The *Contractor* recently notified an early warning as, due to their delay, they anticipate missing the Key Date. How and when do we value this?

This is assessed when the *Project Manager* decides that they have not met the condition stated by the Key Date, not before. At this point the *Project Manager* would assess the additional cost to the *Employer* in carrying out work or paying an additional amount to Others in carrying out their work. For example, this may include re-hire of a crane and storing the scanner in a warehouse. This should be cost incurred 'on the same project' so cannot be remote – it has to be directly incurred on the same project. The assessment should also be made within 4 weeks of the date.

27. Quality control of the supply chain (clause 26.2)

We are about to embark on a high-security new build prison and have selected the ECC contract. The choice of main *Contractor* is critical, which we will inevitably have control over during the tender process. Our concern is the Subcontractors and whether they are of the right calibre – what provisions are there in the ECC contract in this respect?

The ECC provides a level of control in this respect. There is no naming or nomination process. Clause 26.1 confirms that whether the work is subcontracted or not – the *Contractor* retains responsibility. The next clause, 26.2, is key to addressing your concern. This requires that, prior to the appointment, the *Contractor* submits the name of each Subcontractor to the *Project Manager* for acceptance. The *Project Manager* can only 'not accept' if they believe it will not allow the *Contractor* to Provide the Works. Reasons would need to be perhaps around health and safety or major concerns about quality assurance systems or vetting of staff. It is irrelevant what was stated during the tender process in terms of Subcontractors – the ECC contract only becomes operable once signed so the *Contractor* will need to submit at this point. Clause 26.2 therefore provides a level of transparency – ultimately it is the *Contractor*'s responsibility. Failure of the *Contractor* to comply with this obligation may result in the *Employer* terminating their obligation to Provide the Works (clause 91.2, R13).

28. Working with Others (clause 27.1)

We have a complex engineering project under the ECC and have placed direct contracts with a number of suppliers. The civil engineering works are being undertaken by a main *Contractor* under Option A. We have a number of direct contracts with suppliers and other *Contractors* with whom we require the main *Contractor* to liaise with and coordinate its design. How does the ECC contract support this?

The other suppliers and *Contractors* that you have direct contracts with are classed as 'Others' under the ECC (clause 11.2(10)).

The contract refers to Others in a number of instances – in direct answer to your question, clause 27.1 is the most relevant – this requires the *Contractor* to obtain approval of its design from Others as necessary. It is also worth reviewing all of the clauses that refer to Others – these include clauses 16.2, 25.1, 45.2, 91.3, 80.1 and 60.1(5).

 **Some further thoughts and top tips**

- The Works Information is a key document, which the *Contractor* is required to comply with. Make sure this is well understood and implemented.
- Ensure that the design submission and acceptance procedures are fully appreciated at the beginning of the project.
- Make sure that Subcontractors are accepted at commencement of the project (before their appointment is made).
- The Works Information should clearly define the *Contractor*'s interaction with Others – on complex projects this is vital to define accurately.
- The impact of missing a Key Date is significant – ensure these are well planned and the Condition to be met is clearly defined.

Chapter 3
Time

29. Can a Completion certificate also be a payment certificate? (clause 30)

Q We have used the ECSC on a small job that was completed a few months ago. We issued a Completion certificate (which was also the penultimate payment certificate) releasing half of the retention being held. The *Contractor*'s bond people will not accept it as a Completion certificate for that job, making it difficult for the *Contractor* to get another bond for another job. Can a Completion certificate also be a payment certificate?

A There is no such thing as a payment certificate in the ECSC, instead the *Contractor* makes the assessment and the *Employer* decides what is due and pays – see clauses 50 and 51. The requirement for a certificate for Completion to be issued by the *Employer* is set out in clause 30.3. It does not have to be in any particular form, but it does have to state clearly on what date the *works* were completed, and should be issued within one week of that date. It cannot also be a payment certificate because, as said, there is no such thing as a payment certificate. Note in the ECSC there is no separate certifier as such, although the *Employer* can delegate any of their actions to anybody, as long as that delegation complies with clauses 14.4 and 13.1. Whether or not this is accepted by anybody for any reason is not something we are able to comment on.

30. Has Completion been achieved? (clause 30.2)

Our query concerns certifying Completion within an ECC project (Project A), which consists mainly of a large-diameter water main, which the *Employer* currently has beneficial use of. In accordance with clause 11.2(2), all the works within the Works Information have been completed. However, recently a fitting on the water main (which was manufactured, supplied and installed by the manufacturer and/or their subsidiaries) started to leak. This necessitated the new water main being temporarily taken out of service while the fitting and surrounding section was cut out and replaced with a different product. After the repair to this Defect the water main went back into service and the *Employer* gained beneficial use of same again. The faulty fitting had been awaiting transportation to an independent testing laboratory for inspection, analysis and a report into the likely causes and mode of failure. However, another fitting (again supplied, installed etc. by the same company) started leaking on another project (Project B – same *Employer*, different *Contractor*). This fitting has now also been removed and sent for testing and a further report. As you can imagine, the failure of the same fitting on two different contracts has led to some cause for concern within the *Employer*'s organisation as to the potential for further failures and the possibility of further Defects. Contractually, how do things stand within Project A? Obviously the findings of the independent laboratory and report are the immediate concern; however, the *Contractor* may claim that a Defect occurred and in accordance with clause 43, it was corrected. Therefore, while there are no further leaks (Defects) is the *Contractor* now entitled to Completion being certified?

It appears that the *Contractor* has been entitled to have Completion certified some time ago. The *Project Manager* should have certified Completion within one week of the *works* complying with the definition of Completion and his failure to do so is a breach of the contract (clause 30.2). These matters may be Defects, which can and should be, or rather should have been, dealt with by the processes set out in clauses 43, 44 and 45. In addition, as soon as the *Employer* started using the water main or within 2 weeks of Completion (whichever happened earlier), the *Project Manager* should have certified take over – see clause 35. Again, failure to do so within one week is also a breach of contract.

31. What goes onto the first Accepted Programme? (clause 31)

My query is in relation to first Accepted Programme. We have a tender programme that we believe is included as a contract document but not referenced within Contract Data part two. The contract award was delayed by 12 weeks and therefore the tender programme within the contract did not reflect the later contract dates. At contract award Contract Data part two stated 'To be submitted by the *Contractor* in accordance with the contract' (i.e. within 3 weeks). Following contract award we have submitted a first programme for acceptance. This programme was the tender programme adjusted by the delay to contract award but in line with the contract dates and original logic. This programme has not been accepted for the following reasons:

- the *Contractor*'s plans, which it shows are not practicable and
- it does not show the information that this contract requires.

The main issue relates to the *starting date*, which is currently in delay due to *Employer* planning approvals problems. The *Project Manager* is looking for us to submit a first programme, which reflects changes that have occurred since the Contract Date. The *Project Manager* appears to be looking for us to change our logic from the tender programme and show activities happening later as a result of the *Employer*'s delays.

Our query is whether this is reasonable or should we be able to submit a programme in line with our original logic irrespective of changes that have happened since the Contract Date? As there are also a number of *Project Manager*'s instructions already, which are going to be compensation events, any other situation would mean that the programme does not reflect the contract and the baseline at the outset. It should be noted that the programme that was submitted for acceptance did also lack resources and other matters, which we realise need to be added for acceptance. We would be grateful for any guidance on this matter prior to us arranging a risk reduction meeting with the *Project Manager*.

Your tender programme is only the first Accepted Programme if it is referred to in Contract Data part two. Otherwise it is not the first Accepted Programme and we cannot comment on the status of it with such limited information; the incorporation of a programme into the contract could have quite dramatic implications.

Putting this to one side, you need to produce a compliant programme that reflects things as of the Contract Date; this will line up with the other dates and your obligations. Things that happen after that, *Project Manager* instructions to change the Works Information, other *Employer* delays and so on, will be dealt with in the order they are known about and not on the first Accepted Programme. You could not predict these sorts of problems, nor of course does the contract expect you to.

32. Incorporating a programme into Subcontract Works Information (clause 31)

We are using ECS main Option B. The Subcontractor is providing temporary traffic management and the main item is to provide 4 km single lane overnight closures on a dual carriageway. The quantity of closures installed has increased. The Subcontractor has notified a compensation event requesting an enhanced rate/change to the original Bill of Quantities (BoQ) rate. The Subcontractor is using clause 60.1(1) as the basis of his entitlement, saying that any deviation from the prescribed programme and resources therefore constitutes a change to the Subcontract Works Information. The reason for this is that the *Contractor* has identified within the Form of Agreement the appendix where Subcontract Works Information is contained and within this appendix we have included a programme and histogram. Within the Subcontract Data part two we have 'If a programme is to be identified in the Subcontract Data – the programme identified in the Subcontract Data is identified in Subcontract Works Information Appendix A item 4'.

Can the Subcontractor rely on the inclusion of the two documents, the programme and histogram, within the appendix as Subcontract Works Information? And as such is he entitled to a compensation event under clause 60.1(1)? We consider that a change to Subcontract Works Information at the risk of the Subcontractor or a consequence of his proposals is not a compensation event. During the currency of the Subcontract Works the Subcontractor failed to submit a revised programme as required under clause 32. Does this further impact on the Subcontractor's entitlement to a compensation event as he has failed to identify the effects of the compensation event?

Finally, the Subcontractor has only recently identified his entitlement to the compensation event pursuant to clauses 60.1(1), which is several months after Completion. He had previously claimed an entitlement under clause 60.4 but was unable to demonstrate the Defined Cost of his resources had changed. Does this change in argument/direction have the same effect of clause 61.7 and effectively time bar any compensation event claim? Note that the subcontract had sectional Completion Dates and as such certain sections are now beyond the *defects date*.

If you have somehow included a programme within the Subcontract Works Information then if you change that programme that becomes a compensation event because you are changing the Subcontract Works Information. And that compensation event will be valued not based on the rates in the BoQ, but on the actual increase in Defined Cost. So it is not a case of enhancing a rate, the Subcontractor has to demonstrate the Defined Cost. That may be more or it may be less than he has in the BoQ, but he still has to demonstrate it. It is the Subcontractor who is supposed to provide the programme, not you. It is for you to accept or reject it. It is his programme for his *subcontract works*. We suggest you look at clauses 11.2(1) and 31 and 32 for more information on this, and it is in the Subcontract Data rather than the Subcontract Works Information because programmes will inevitably change! So your mistake was saying at the entry in the Subcontract Data – look in the Subcontract Works Information for the programme. Because now, whenever it changes, it is a compensation event. The time bar in clause 61.3 does not apply to clause 60.1(1) because it is a compensation event that you were supposed to notify, not the Subcontractor – see clause 61.1. We cannot see how the time bar in 61.7 can apply when you have not yet got near the *defects date*. The answer to that will be in what you have set the *defects date* up to be in the Subcontract Data. So, yes it is a compensation event, but the Subcontractor still has to demonstrate what effect it had on Defined Cost, just as they do with any other compensation event.

33. Programme float and time risk allowances (clause 31)

On an ECC Option A contract, in presenting the clause 31 programmes, the *Contractor* has not defined any float or time risk allowance; rather it has included a terminal risk allowance. With particular reference to the critical path, we consider that not only does this preclude the ability to monitor risk/residual risk as the works progress, but it also prevents assessment in terms of the sufficiency of the operational resource and time allowances and the deliverability of the programme generally. It is worth noting that the critical path constitutes complex bridge construction works over an environmentally sensitive water course. Could you advise on the resolution of the issue, with regard to the *Contractor*'s position that there is no float within the critical path, and an assessment that a programme with no float on the critical path is impractical/unrealistic and, therefore, cannot be accepted?

A risk reduction meeting cannot fully resolve the matter (the *Contractor* has already stated that, due to its commercial approach on the job, there is no float/risk provision in the critical path operations) and there only appears to remain a completely unsatisfactory situation whereby these mutually exclusive positions lead to an impasse undermining the programme as a management tool.

The *Contractor* is required to show float and time risk allowance on its programme – see clause 31.2. If the *Contractor* does not do so, then the *Project Manager* can notify his reasons for not accepting under clause 31.3, second bullet. A programme without any time risk allowance or float, or where all the time risk allowance is at the end, is also simply not practicable to achieve and should therefore not be accepted under the first bullet of clause 31.3.

However with regard to the critical path, the *Contractor* is absolutely correct that there simply cannot be any float within that – by the very definition it is the path that is critical. In fact, one of the accepted definitions of the critical path is that path that has zero float, or, to put it another way, it is the quickest period that the work can be carried out in. Generally, now, computer software will define the critical path and float for the remaining items once the programmer has entered the time periods for each activity (including any time risk allowances for that activity) and the links between the activities (either logic or resource driven). However, the *Contractor* is not best advised to lump all its time risk allowances together at the end of the programme. These should be included for each activity that needs them (not all will). The amount of these will depend on many factors including the nature of the activity and the time of year it is being carried out. That way the programme shown will be achievable (i.e. practicable). Time risk allowances remain the *Contractor*'s and the *Employer* cannot use them when assessing the effects of a compensation event.

34. What becomes of a tender programme? (clause 31)

As part of a quality assessment, we asked tenderers for a resourced programme submitted from the stages of design to Completion. As you might expect, these varied in quality. Contract Data part one stated that 'The *Contractor* is to submit a first programme for acceptance within 4 weeks of the Contract Date'. We have a few questions on this scenario:

1. In awarding the contract, are we bound to accept the tender programme as the clause 31 programme?
2. Are we entitled to request a new clause 31 programme based on the actual Contract Date?
3. At award of contract, the Contract Date was 2 weeks later than had been assumed by the tender programme. Is the *Contractor* entitled to have Key Dates and/or the Completion Date moved forward by 2 weeks?

The answer to question 1 is going to depend on the status of the programme issued at tender stage. If it was referred to by the *Contractor* in the Contract Data part two then it becomes the first Accepted Programme – see clause 11.2(1); the reference in Contract Data part one is irrelevant because the 'if' statement does not then apply.

With regard to question 2, we are not sure that in practice this is particularly relevant, because the programme has to be revised in any event at regular intervals – see clause 32.

With regard to question 3 the Contract Date has no relevance to a programme issued for acceptance – it is the *starting date* and *access dates* that are important – see clause 31.2. These are the dates that were in your Contract Data, unless these were perhaps amended in tender correspondence.

35. What do we do if the programme shows a wrong date? (clause 31)

On an ECC contract, the *Contractor* is asking for acceptance of a programme that shows the effects of a *Project Manager*'s instruction, which is a compensation event affecting both the Prices and planned Completion. I believe the *Contractor* has not assessed this correctly and the assessment is not currently agreed between us. I have not accepted the submitted programme because the programme shows the *Contractor*'s critical path delay of 53 days, whereas my assessment of this is 29 days. The programme thus indicates an incorrectly revised Completion Date. Is this a reason for not accepting under clause 31.3?

I am looking at the second bullet here and considering the direction of clause 32.1 second bullet and clause 32.1 first bullet. I do not understand how I can accept a programme that shows something that is not agreed. This compensation event is significant in value and I believe will take some time to settle (maybe by means of the *tribunal*) as we have reached an impasse.

Subsequent to this event, we have had a further significant compensation event involving a delay, which is agreed in both cost and time. I am therefore required to accept a further programme to reflect this change. How should I progress this if the previous matter is not agreed?

There are two separate dates that have to be shown on the programme for acceptance. The date for planned Completion is one such date – see second bullet of clause 31.2. This is the date on which the *Contractor* plans to achieve Completion. It will be at the end of the critical path and all of the activities will, directly or indirectly, link to it. It can be before, on, or after the Completion Date. If it is the latter the *Contractor* is admitting it is going to finish 'late'.

The Completion Date is the other such date – see the first bullet of clause 31.2. This is the date by which Completion should be achieved (see clause 31.1). This is not linked to anything else on the programme and can only be moved as a result of a compensation event (clause 63.3) or agreed acceleration (clause 36.3).

You need to make sure you understand the difference between these two and do not confuse them. You also need to understand that terminal float (i.e. the period between planned Completion and the Completion Date, remains the *Contractor*'s responsibility (plus or minus) and cannot be used to alleviate the effect of a compensation event – see clause 63.3).

With regard to the compensation event, the time for 'settlement' is irrelevant to you. If you are not happy with the *Contractor*'s quotation, clauses 64.1 and 62.3 require that you make your own assessment of both the money and time consequences of the compensation event. Clause 64.3 requires that you do that within 3 weeks. Once you notify the *Contractor* of this assessment the compensation event is implemented under clause 65.1. You then include in your payment certificate the amount you have assessed (Options A or B) when these works are complete or add it to the target (the Prices, in Options C or D). You also extend the Completion Date by the amount you consider it is due. That is the end of your involvement.

The *Contractor* can (possibly negotiate with the *Employer* or) go to adjudication to get it changed. If the *Contractor* believes it will take 53 days to deal with this event it is entitled to put that on its programme for acceptance, and therefore move the date for planned Completion by that amount. That will not move the Completion Date and if you accept that programme it will not mean you are accepting that the Completion Date should move by that, or any other, amount. On the other hand if you do not think it will take 53 days to deal with the event, then you can reject the *Contractor*'s programme because it does not represent the *Contractor*'s plans realistically (clause 31.3, third bullet). Of course, if you are wrong on that then eventually the *Contractor* will be likely to win the argument if it goes to adjudication.

36. Problems with a programme submitted at tender stage (clause 32)

We are at present engaged on a major construction project using the ECC Option A and seek clarification on a point of procedure regarding the programme. All tenderers submitted a programme as part of their tender, as they were required to do. Following evaluation of the tenders the *Employer*'s agent (this role includes acting as *Project Manager*) recommended award to *Contractor* X. No comment was made at the time by the *Employer*/agent regarding *Contractor* X's programme and there was no notification of acceptance or reasons for non-acceptance of this programme. The contract was formally awarded by the *Employer* to *Contractor* X (the *Contractor*) in early October.

The *Project Manager* advises that the *Contractor* submitted its first programme for acceptance in January. The *Project Manager* wrote to the *Contractor* 13 days later with comments that required attention prior to the *Project Manager*'s acceptance. The *Contractor* responded claiming that the *Project Manager*'s response had neither accepted nor notified reasons for not accepting, stating that one of these should have been due within 2 weeks of submission and they therefore deemed the programme submitted in January to be the Accepted Programme. The *Contractor* then submitted an updated programme in February taking cognisance of the comments and requirements sent by the *Project Manager* earlier. The *Project Manager* wrote to the *Contractor* more than 2 weeks later indicating that the programme submitted in February was not accepted for all four clause 31.3 reasons. The *Contractor* maintains that there is an Accepted Programme, probably referring to its 'deemed' Accepted Programme.

The ECC guidance notes relating to clause 50.3 say that if a programme was required to be submitted with the tender, the programme is identified in the Contract Data at the Contract Date. In this situation an Accepted Programme already exists and no amount can be retained under this clause. Managing Reality Book 3 says that 'once the programme has been accepted, it becomes the Accepted Programme'. Managing Reality Book 5 says that, 'where the programme is submitted with the tender, the *Project Manager* should inform the *Contractor* whether he has accepted the programme or not before the execution of the contract'. Brian Eggleston in his commentary on the NEC3 says that 'the contract does not deal expressly with the position if the *Project Manager* finds the programme identified in the contract data unacceptable. It is by definition the Accepted Programme. Presumably this is a matter to be dealt with before the award of the contract.'

None of the above considers the possibility in which the *Project Manager* neither accepts nor notifies reasons for not accepting, as appears to be the case on our project. We would welcome any comments and advice you may be able to offer, particularly regarding whether an Accepted Programme (or deemed Accepted Programme) exists and to which one would this relate. Also, if this relates to the one submitted with the tender, how should any deficiencies in information that would have led to it not being accepted had it been properly assessed at the time, be dealt with now?

The first thing you and your *Contractor* need to understand is that the Accepted Programme is not a single programme; it is a continually evolving one. The contract requires the *Contractor* to submit revised programmes at regular intervals – note the final bullet of clause 32.2 and the entry you have in your Contract Data part one. We always recommend that the interval should be no longer than monthly, ideally shorter. Once each programme has been accepted, it becomes the Accepted Programme until the next one is issued and accepted (clause 11.2(1)). And the cycle goes on.

The status of the programme submitted with the *Contractor*'s tender depends on how it has been referred to within the Contract Data, so unfortunately we disagree with one of the quoted suggestions above that its status is unclear – nothing could be clearer. If the programme has been referenced in the Contract Data part two by the *Contractor* then it automatically becomes the first Accepted Programme – the wording of clause 11.2(1) is very clear on that. Of course, it then gets superseded by future programmes, once they are accepted. If it is not referred to in the Contract Data it is not an Accepted Programme. If the tender assessment team are not happy with this programme, then do not accept the tender, or put up with it or negotiate (if procurement rules permit).

Looking at what has actually happens it seems to us that there has been a catalogue of errors on both sides. Firstly, it seems that the *Contractor* was very late in submitting its first post-contract programme for acceptance. And the *Project Manager* has also not done what he or she should have done. If the tender programme was not the Accepted Programme, for the reasons in the previous paragraph, the *Project Manager* should have been retaining 25% of the Price for Work Done to Date in assessments until a 'correct' programme (clause 50.3) was submitted. Even if it was the Accepted Programme the *Project Manager* should have been chasing for another one, and instructing one if need be (see first bullet of clause 32.2). Once the *Project Manager* receives a programme he or she has 2 weeks to say 'yes' or 'no', and if he or she says 'no' he or she has to give reasons (clause 31.3). If the *Project Manager* has some comments he or she wants to discuss with the *Contractor* it is a bit too late to write after 13 days. Why not just talk to the *Contractor* about the programme and discuss any concerns? And then the *Project Manager* failed to reply within the 2 weeks, which is a compensation event in itself (clause 60.1(6)). But then the *Contractor* got it wrong, as well. There is no such thing as a 'deemed accepted' programme. As previously mentioned, the *Contractor*'s protection is through the notifying of a compensation event. But it is now too late to do that – see clause 61.3. The *Contractor* then submitted a new programme for acceptance – absolutely correct. But the *Project Manager* has now not accepted that for all four reasons in the contract. That means it is either an appalling programme or the *Project Manager* is perhaps covering himself or herself against all eventualities, or somewhere in between. None of those options bodes well for the future.

The present position is that you either have no Accepted Programme at all or the (well out of date) one in the Contract Data is the Accepted Programme. That can be in neither Party's interest. So our first piece of advice is for the *Project Manager* to sit down and talk constructively with the *Contractor*. Neither has covered themselves in glory so far, so both should approach those discussions with a degree of humility, and, above all, with a positive desire to get a programme that both sides can live with. That is far better than having no Accepted Programme at all, or worse still, an out of date, and therefore useless, one. And, in any event, acceptance does not impose any liability on the *Employer* or the *Project Manager* – see clause 14.1. A further point is that until the *Contractor* starts issuing programmes for acceptance within the periods in the contract, and the *Project Manager* accepts them otherwise the *Project Manager* is required to assess all compensation events himself or herself, using what he or she believes the Accepted Programme should be – see clauses 64.1 and 64.2, and he or she has to do that within 3 weeks – see clauses 64.3 and 62.3.

37. Lack of acceptance of programmes (clause 32)

We are a Subcontractor on a contract using the ECS. We are struggling to get acceptance of any of our programmes and have to date submitted 10 programmes for acceptance by the *Contractor*, of which all 10 have not been accepted. Each time our programme is not accepted the *Contractor* changes the *subcontract access dates* and we resubmit the programme with the new access dates for some sections of the work and again it is not accepted. The work is progressing ahead of our original programme. Can this situation carry on until the end of the project where the *Contractor* keeps on not accepting our programme, or is there a mechanism in the ECS where they have to accept a programme?

You are required to submit programmes for acceptance as set out in clause 32.2. Once you have done so the *Contractor* has 2 weeks to accept or not accept it (clause 31.3). If the *Contractor* does not do so within that period then that is a compensation event (clause 60.1(6)), and you will be compensated for the effect the event has had on the Defined Cost and the programme. If the *Contractor* does not accept it, the *Contractor* has to give reasons (clauses 31.3 and 13.4). You are required to resubmit the programme taking into account those reasons (clause 13.4). If the reason is listed in the contract (see clause 31.3 for those) then that is the end of the matter (clause 13.8). If the reason is not one of those listed in the contract then that is also a compensation event (clause 60.1(9)). The problem you may have with these compensation events is that you have only got 7 weeks from when you knew about them to notify them as a compensation event, otherwise you lose all your rights to additional time and money – see clause 61.3. In addition, if the *Contractor* issues an instruction changing the *subcontract access dates* (as set out in the Subcontract Data) each of those instructions will be a compensation event (clause 60.1(2)).

38. Showing non-implemented compensation events on a programme (clause 32)

Under an NEC3 ECC Option B contract, we have a number of instances in which the *Project Manager* has instructed a change to the Works Information but the compensation event has yet to be implemented (as the quotation is either outstanding or not yet agreed); the effects on the programme have therefore not been accepted. The *Contractor* has submitted its latest clause 32 programme for acceptance and included the programme effects for these compensation events, which are not yet implemented.

As the *Employer*, we view that this is incorrect as we have not yet notified acceptance of these compensation events and indeed clause 32.1 only mentions that the effects of implemented compensation events be included within the programme. Is this position correct?

We are concerned that if we do accept a programme that includes the effects of implemented compensation events we may by default be agreeing to the time effect of the compensation event when this is not the case. However, we appreciate that the *Contractor* is keen to reflect the actual position within their programme with regard to instructed work that has been completed before the compensation event has been implemented – thus to ensure they show this work as as-built on the programme.

It is important to understand that a programme submitted for acceptance has to show both the Completion Date (contractual) and planned Completion (practical) – see clause 31.2. The latter is linked through the activities in the programme, the former is not. The latter can move for all sorts of reasons, the former can only move because of a compensation event. If the *Contractor* knows something is going to happen, it has to show its effects on the programme straight away, whether it is a compensation event or not, and that may or may not affect planned Completion. That is because it has to show 'the order and timing of the operations which the *Contractor* plans to do...' (third bullet of clause 31.2).

The practical effects of events, therefore, including compensation events that are known about but not yet implemented, have to be shown on the programme. So planned Completion may move, but the Completion Date does not and cannot at that stage. Once the compensation event is implemented then the Completion Date is moved and that movement is shown on the programme. As long as the *Contractor*'s programme follows those rules, then the *Project Manager* should accept it, as long as you are happy with it in other respects. Acceptance will not infer or imply anything about the compensation event. And it will not take any responsibility from the *Contractor* – see clause 14.1.

39. Take over and effect on liability (clause 35)

What effect does certifying take over for part of the *works* have on the *Contractor*'s insurances/liability for that part. We have an ECC Option A contract for environmental improvement works (new footways, lighting, etc.) in which vandalism and anti-social behaviour have increased in recent months. As a result, the *Contractor* is keen to protect himself at the earliest opportunity.

Take over of all or any part of the *works* occurs when the *Employer* starts using that part (or all), unless that use is for a reason stated in the Works Information or to suit the *Contractor*'s method of working – see clause 35.2. One of the *Employer*'s risks listed in clause 80.1 is loss of or wear or damage to any part of the *works* that have been taken over (fourth main bullet). Up until take over the *Contractor* is responsible for this (clause 81.1) and has to insure it (clause 84.2). Take over is therefore important because once any part of the *works* is taken over, the *Contractor* is not liable for damage unless they cause it.

40. An *Employer*'s risk? (clause 35)

Q I am the *Project Manager* on a project using ECC3 Option C. Completion was achieved on 4 June and the *Employer* took over the works on 11 June. However, a few days later, due to the heavy rain, topsoil was washed off an area of the embankment slope, which also took with it landscape planting. The *Contractor* states the scheme is handed over. We state it is a Defect as the topsoil layer should not have failed.

A The first place we would look in the contract is the set of *Employer*'s risks in clause 80.1. The fourth main bullet says:

'loss of or wear or damage to the parts of the *works* taken over by the *Employer*, except loss, wear or damage occurring before the issue of the Defects Certificate which is due to ... a Defect which existed at take over ... or ... the activities of the *Contractor* on the Site after take over'.

If it was the case that there was no Defect that existed at take over, or the *Contractor* has carried out no untoward activities after take over to cause this, then this matter would be an *Employer*'s risk. The *Contractor* would have done that which it was obliged to do, which is to Provide the Works in accordance with the Works Information.

A Defect is a part of the *works* that is not in accordance with the Works Information (clause 11.2(5)). Were tests or inspections carried out to verify whether the *Contractor* had done that which the Works Information stated it should do? Was the Works Information sufficient in itself – that is, did the *Contractor* do right but the brief was wrong in the first place? How can you now be sure that there was indeed a Defect in how the *Contractor* had provided the *works*? How can you be sure the constructed works should not have failed?

We would imagine it is going to be extremely difficult to prove a Defect existed at the time of take over. If somehow it can be proved, then this would be a *Contractor*'s risk under clause 81.1. Are there other areas of similar topsoil and planting that demonstrate the Works Information has not been followed? Maybe the Works Information was correct, the *Contractor*'s work was acceptable, but it was unfortunate timing that this part of the *works* had not had a chance to settle before the damaging rain – this is not a *Contractor*'s risk after take over.

41. Damage by Others after take over (clause 35.2)

We are using ECC Option C and would like to know what is the correct (contractual) position in respect of damage caused by Others? For example, in the instance that we, as the *Contractor*, had completed our *works* but had since handed over an area of the completed *works* to another *Contractor* (who is not part of the contract) by request of the *Project Manager*. What is the position of the *Contractor* when the *Project Manager* states that the *Contractor* was principal *Contractor* and therefore holds responsibility for damage caused by others (irrespective of the handover documentation that was issued to the other *Contractor* from the principal *Contractor* clarifying the state of the completed *works*). You should note that the contract makes no provision in allowance for interim handovers to Others and that there is no provision or allowance for resources in the Prices for monitoring those works carried out by others.

We are afraid that there is no such thing as 'handover' in the ECC so it is difficult to comment on this, other than to reflect on the correct wording/processes in the ECC. NEC contracts are not the same as other contracts and all need to take care to use the correct principles and terms when managing them or confusion might arise.

If the *Employer* wants to use a part (or all) of the *works*, either for their work or for the work of Others (as here), then they take over the *works* – see clause 35.2 (we assume from what you say that neither of the two exceptions in the bullets applies in this case). The *Project Manager* is then required to certify the date for take over within one week, see clause 35.3. That is what you should have got him or her to do, rather than using the term handover.

The fourth main bullet of clause 80.1 makes it clear that the *Employer* is at risk for damage to the *works* once they are taken over, not the *Contractor* (see clause 81.1). And you are no longer required to insure the *works*, because the first sentence of clause 84.2 makes it clear that the *Contractor* only has to insure against its risks. The fact that you are principal *Contractor* is irrelevant. The CDM Regulations specifically state they impose no civil liability of any sort. And in any event the contract is clear as to whose risk it is (i.e. the *Employer*'s).

42. Where use of the *works* may not constitute take over (clause 35.2)

On an ECC Option A contract the *Employer* wishes to use a part of the *works* for a number of weeks but considers that the *works* are not complete and believes and is concerned that such use will automatically trigger Completion for that part of the *works*, the beginning of the countdown to the *defects date* for that part of the *works* as well as having insurance consequences for the *Employer*. To that end the *Employer* has requested the *Contractor* to confirm in writing that such use does not constitute take over. Such confirmation is required by the *Employer* as a pre-condition of the *Employer* using the *works*. If the *Contractor* does not provide such confirmation, the *Employer* will be obliged to use an area that is outside the Site.

At the request of the *Employer*, a 'Statement of Readiness of Use' has also been provided by the *Contractor* as a second pre-condition of the *Employer* using the *works*. The statement provided confirms that the *works* are complete but for two items, which in no way prevent or hinder the immediate and full use of the *works* as requested by the *Employer*.

The *Contractor*'s view is that such use of the *works* by the *Employer* does constitute take over of that part of the *works* under the contract, but that that does not automatically equate to Completion either of the whole of the *works* or, as the *Employer* believes, of that part of the *works*. The *Contractor* does not dispute that they must complete the *works* whether or not take over has been certified. Note that clause 11(2) has been amended to:

> 'Completion means in any event a state in which the Works are complete in all respects and free from apparent Defects, save for any minor items of incomplete works or minor Defects the existence, Completion or rectification of which in the opinion of the Project Manager would not prevent or interfere with the use and enjoyment of the Works; provided where it is expressly stated in any provisions of the Works Information that the testing, commissioning, regulation or adjustment of any mechanical or electrical services is to be completed before practical Completion of the works shall not be considered to be practically completed until the same is done as the Contract Documents require.'

The *Employer* appears to have misunderstood the contract. Completion and take over are two entirely different concepts in the ECC. Completion as about what state the *works* are in – that is, do they meet the requirements set out in clause 11.2(2)? Completion is what the *Contractor* has to achieve before the Completion Date. Take over is about who is using the *works*, as set out in clause 35. Once the *Employer* starts using the *works* the *Employer* takes them over. The *Employer* is entitled to start using all or any part of the *works* before Completion, and if they do so they have taken over those parts of the *works*, see clause 35.2. However, Completion will still not be achieved until the *works* have met the tests set out in clause 11.2(2). At take over the *Employer* becomes at risk for any loss of or wear or damage to the *works* – see the fourth main bullet of clause 80.1. As these are no longer at the risk of the *Contractor*, the *Contractor* is no longer required to insure them, see the first sentence of clause 84.2. However, the *defects date* is related to the date of Completion, not the date of take over, see the relevant entry in the Contract Data part one. If the *Employer* does take over early, i.e. before Completion and before the contractual Completion Date, that will be a compensation event. However, if the *Employer* takes over in the situation in which the *Contractor* is already late, i.e. before Completion, but after the Completion Date, that is not a compensation event. See the wording of clause 11.2(15). It is for the *Project Manager*, not the *Contractor*, to decide when Completion and take over has occurred and certify them, see clauses 30.2 and 35.3. The *Project Manager* does so by applying the meaning to those terms set out in the contract. There is no such thing as a certificate of readiness of use in the ECC; you are not required to provide it and it means nothing.

As a final point it appears to us that the *Employer*'s changes to clause 11.2(2) are a recipe for an argument and are totally subjective. The drafting seems to introduce new defined terms and does not properly use the identified terms. What exactly is a 'minor' item of work? And what does the reference to practical Completion mean? The contract was designed to remove subjective judgements such as these, and replace them with objective decision making, by defining what is meant by Completion. This change has re-introduced the uncertainty that vague words such a 'substantial Completion' and 'practical Completion' bring.

43. The interaction of take over, Completion and delay damages (clause 35.2)

We are currently engaged on an ECC Option A project including Option X5 and X7. Option X5 defines 27 individual Completion Dates and Option X7 allocates delay damages to 26 of the 27 Completion Dates.

1. The Works Information on this project is silent on what constitutes Completion for any of the individual Completion Dates and a dispute has arisen over when Completion was achieved. As the *Contractor*, we believe that Completion was achieved at latest when the *Employer* provided access to Others (*Contractors*) to start their works. Is the *Contractor* correct with this view in the absence of any other Completion criteria in the Works Information?

2. A number of the individual Completion Dates are based on providing a complete set of eight bases. The *Employer* has started erection works prior to all eight bases being complete but has refused to provide Completion for the bases that it has effectively taken over early. How should the *Project Manager* have dealt with this situation?

3. What are the obligations of the *Project Manager* regarding deduction of delay damages? Is the *Project Manager* obliged to deduct delay damages in full unless the Completion Date(s) are altered by agreed compensation event? What happens with delay damages in a scenario such as question 2 above, should the *Project Manager* levy delay damages in full?

4. If the *Project Manager* fails to deduct delay damages from an interim assessment is the *Project Manager/Employer* time-barred at any stage from subsequently revisiting the applicability of delay damages? We have received notification from the *Project Manager* requesting repayment of delay damages because they were not deducted in every monthly assessment they should have been.

5. The wording used in Option X5 for each of the Completion items is slightly different from the same headings used for the delay damages under Option X7, for example under X5 the Completion Date is defined as a date for Array A but under Option X7 it is defined as a sum for Array 1. Is it imperative that the wording in X5 and X7 match for the provisions on interim Completion and delay damages to apply?

Completion of a *section* of the *works* is dealt with in exactly the same way as Completion (see clause X5.1). In the absence of anything in the Works Information, the definition of Completion is set out in the final part of clause 11.2(2). It is when 'the *Contractor* has done all the works necessary for the *Employer* to use the works and for Others to do their works'. It appears the *Contractor* has achieved that by the fact that Others (i.e. these other *Contractors*) are carrying out their works.

However, you must also remember as well as Completion there is also take over to consider. Completion is about the state the *works* are in (clause 11.2(2)), whereas take over is about who is using the *works*. The *Employer* takes over the *works* (or any part of the *works*) when it starts using them (clause 35.2), and that would include getting Others to use them. The *Project Manager* has to certify take over as well, within one week of the use starting (clause 35.3). So, on the information you have given us, take over must have occurred, regardless of whether or not Completion has been achieved, and the *Project Manager* is in breach by not certifying as such.

You will see in X7 that delay damages cease on the earlier of Completion or take over happening. Therefore, delay damages cannot be applied to these *sections*.

In addition, the *Employer* can take over part of the *works* or part of a *section* of the *works* at any time. For example, in your case, the *Employer* could start using six out of the eight bases in a *section* (clause 35.2), in which case those six bases are then taken over. In that case, the delay damages for that section are reduced in the ratio to the amount of benefit the *Employer* has got from the six as opposed to the eight – see clause X7.3.

The *Project Manager* (not the *Employer*) deals with the deduction of delay damages as part of his or her assessment – see the third bullet of clause 50.2. The *Project Manager* is obliged to deduct them when they become payable. If the *Project Manager* does not do so, that is an error, but this can be corrected in any later assessment – see clause 50.5. There is no time bar for this to be done.

Obviously the *sections* should have the same name in Options X5 and X7. If they do not that can cause problems, but that is a legal matter, which we are unable to offer advice on. In reality, in the scenario you have set out, it is unlikely that any delay damages would be payable, assuming that they started using the bases by the Completion Date for those bases. The fact that the *Project Manager* has not done what he or she should (i.e. certifying Completion and/or take over) does not mean you are liable.

 **Some further thoughts and top tips**

- Make sure that the state of Completion is properly thought about and defined in the Works Information when preparing the tender documents.
- Think of the programme requirements in clauses 31 and 32 to be a set of minimum requirements – there is nothing to stop including more than the minimum if parties agree it would help the contract management.
- With the first programme in particular, the *Contractor* should get the person(s) who prepared this document to spend some time walking the *Project Manager* through this to understand better the logic and any pinch points that concern the *Contractor*.
- Make sure that take over, testing and Completion are understood, properly reflected in the tender documents, planned out properly and clearly shown on the Accepted Programme.

Chapter 4
Testing and Defects

44. Frustrated *Supervisor*! (clause 40.1)

Q I am a frustrated *Supervisor* under an ECC Option A contract! I was not involved in drafting the original Works Information but understand that no specific tests or inspections were stated. What tests and inspections should the *Contractor* undertake and how can I instruct additional tests and inspections?

A The frustration is appreciated and, worryingly, not uncommon. Drafting the Works Information correctly is vital and should be a team effort – taking on board all expertise.

If the Works Information is silent then the only tests and inspections that the *Contractor* is required to undertake are those required by the applicable law (clause 40.1).

If you suspect a Defect you can as *Supervisor* instruct the *Contractor* to search. However, if there is no Defect then this will be a compensation event (clause 60.1(10)).

It sounds like you are better off redrafting the Works Information and then getting the *Project Manager* to change the Works Information (clause 14.3). This, however, will be a compensation event (clause 60.1(1)). It may be more sensible to get the *Project Manager* to notify an early warning and discuss the impact of the additional tests and inspections you now require – a view can then be taken.

Ultimately, if the work does not comply with the quality standards within the Works Information then it will be a Defect (clause 11.2(5)).

Recent guidance published by the NEC provides clear direction on how to draft the ECC Works Information.

45. What does 'unnecessary' mean? (clause 40.5)

 I am trying to understand what 'unnecessary' means under clause 40.5 of the ECC. I recognise that if a test or inspection by the *Supervisor* causes unnecessary delay then this would be a compensation event (see clause 60.1(10)), but what does this mean?

 It is a good question and one that could be subjective. Our advice would be to define the time periods within the Works Information. Tests and inspections normally fall into three categories: tests and inspections that the *Contractor* undertakes, those undertaken by the *Supervisor* and those witnessed by the *Supervisor*. Why not create a table in the Works Information and, against those that involve the *Supervisor*, include a timescale. The NEC published guidance on how to draft the Works Information in April 2013. Section WI700 would be the appropriate section to define this.

46. Charging costs incurred for repeating tests and inspections (clause 40.6)

As an *Employer* we have engaged a consultant *Supervisor* on an hourly rate to manage our ECC project. We are having a number of problems with our *Contractor* and numerous tests need to be repeated as a result of Defects being found when tests are undertaken. Are we allowed to charge the *Contractor* for this?

Yes, clause 40.6 allows the *Project Manager* to assess the cost to the *Employer* in repeating the test or inspection. The *Contractor* pays the amount. While the work is underway the cost will be deducted from the amount due (second bullet of clause 50.1).

47. Is there a snagging list? (clause 42.2)

Reading clause 42.2, in conjunction with clause 13.7 it would imply that every single Defect should be notified separately – surely this cannot be right as we have hundreds of Defects on our project?

There is no ambiguity. Clause 42.2 requires that until the *defects date* the *Supervisor* and *Contractor* notify one another of each Defect. The 'notification' needs to be undertaken separately as per clause 13.7, as you say. Each one may have a differing *defect correction period* depending on how these have been completed in the Contract Data so it is important that they are separate.

This rigour ensures that we overcome the problems of old – poor and ambiguous paperwork. Perhaps there will be more paperwork but it will be very clear where we stand with each and every Defect. The NEC published guidance on the use of communication pro formas in April 2013. It is useful to agree a communication protocol at the outset of the project and agree how this will be managed.

48. Pro forma for latent Defects (clause 42.2)

 Is there a latent Defects pro forma?

 We are not too sure what 'latent' Defect, latent to what? Assuming you are using an ECC contract, then all Defects (note definition in clause 11.2(5)) found up until the *defects date* are notified by the *Supervisor* to the *Contractor* (see clause 42.2). You can use a generic notification form or you can design your own specifically for this. Strictly speaking, none of these are latent Defects as such. After the *defects date*, any Defect the *Employer* finds, which is not listed on the Defects Certificate, may well, in law, be a latent Defect, but you will need to get legal advice on that. The contract keeps the *Employer*'s rights with regard to these; see the last sentence of clause 43.3. The law will then decide how and when these should be notified. However, in such case there is no automatic right to correction (as there is before the *defects date*), and you will need to take legal advice as to how to proceed.

49. Constraints in the Works Information (clause 42.2)

The *Contractor* has designed a heating system to provide a room temperature at 10°C. The Works Information states this should be 15°C. I, as *Project Manager*, have therefore currently advised the *Contractor* that I am disallowing costs under clause 11.2(25) fifth bullet point as this is a Defect associated with not complying with a constraint in the Works Information. This has been challenged on the basis that the Works Information does not contain a schedule of constraints (i.e. compliance or not with a specification clause is not deemed as a constraint). Do constraints have to be specifically identified in a separate document/highlighted in say a list or can they just be items included within the Works Information?

We assume that the specification is within the Works Information. We consider that you were wrong to treat this as a Disallowed Cost, but not for the reason the *Contractor* suggests. The constraint is on 'how' the *Contractor* Provides the Works, not on 'what' he provides, as is the case here. This only applies to method specifications, not results-based specifications, as this one clearly is.

Such constraints do not have to be in a formal schedule of constraints as is suggested by the *Contractor*, but they do have to be clear and unambiguous. However, that does not apply in your case for the reason given previously. This is, of course, still a Defect that the *Contractor* must correct – see clauses 11.2(5) and 43.1.

The *Supervisor* (not *Project Manager*) should therefore notify it under clause 42.2. The cost of correcting this Defect will be paid, as long as it is corrected before Completion (see fourth bullet of clause 11.2(25)).

50. Dealing with a Defect (clause 43)

On our project the *Employer*'s Works Information specifies some mesh access flooring to be 45 mm thick. The *Contractor* has installed this at 43 mm thick on the ECC Option C contract. The work is underway – how do we best deal with this?

The use of the 43 mm will be a Defect, because it does not comply with the Works Information – see the first bullet of clause 11.2(5). The *Supervisor* (not *Project Manager*) should therefore notify the *Contractor* of this Defect – see clause 42.2. The *Contractor* is then required to correct it under clause 43.1. That has to be carried out before Completion, if it will prevent Completion (see 11.2(2)), otherwise it has to be carried out within the *defect correction period* set out in the Contract Data (see 43.2). If the *Contractor* does not correct this within that period then the *Project Manager* assesses what it will cost the *Employer* to get it corrected by other people and deducts that from the *Contractor* – see clause 45.1. As to payment we note that this is an Option C contract, which shares the risk of Defects between the Parties. The answer will therefore depend on when the correction is carried out. If it is carried out after Completion then the cost of it will be a Disallowed Cost – see the fourth main bullet of clause 11.2(25)). Otherwise the cost forms part of Defined Cost and is paid. In Option C, the risk of the *Contractor*'s errors is, with few exceptions, shared between the Parties.

51. Dealing with Defects after Completion (clause 43)

 We have an ECC project. The project achieved Completion 2 months ago and we are debating with the maintenance team what kind of warranty periods the ECC provides for one particular aspect – the air handling units.

 The 'warranty' period in the ECC is effectively the time period between Completion and the *defects date*. This is typically 52 weeks but it is whatever has been completed in the Contract Data. During this period there is a dual obligation – the *Supervisor* needs to notify the *Contractor* of Defects and the *Contractor* is required to correct within the *defect correction period* (again, defined in the Contract Data).

In addition, even when the Defects Certificate has been issued that does not affect your rights to recover the monies from the *Contractor* for that breach (see the last sentence of clause 43.3).

52. No operational and maintenance manuals! (clause 43)

I undertook *Project Manager* duties on a small ECC contract last year. The *Contractor* has now contacted me for payment of retention money (2.5% outstanding); however, the *Contractor* has not provided the operational and maintenance manuals that were detailed in the Works Information. Should I have notified the *Contractor* that this was a Defect for the non-compliance with the contract? And although I did not, can I still notify a Defect and withhold the retention money until this information is provided? It is my understanding that I can withhold approval of retention money until the Defects Certificate has been issued.

The *Supervisor* should have notified these omissions as Defects as soon as he or she was aware of them. Whether he or she can do so now depends on the *defects date*, which is defined in the Contract Data as being a period after Completion. If the *defects date* has passed it is now too late to notify a Defect – see clause 42.2. In that case all the *Supervisor* can do now is issue the Defects Certificate, certifying all notified Defects have been corrected, see clause 43.3 and also clause 11.2(6). The *Project Manager* is then required to issue the final certificate (see clause 50.1), which will include the release of all retention (see clause X16.2). If the *defects date* has not yet passed, the *Supervisor* (not the *Project Manager*) should immediately notify the *Contractor* that these omissions are a Defect. He does not need to instruct the *Contractor* to correct them because the contract already does that. The *Contractor* then has to correct those Defects within their *defect correction period*, which starts when the *Supervisor* notifies the *Contractor* of the Defects (see clause 43.2). The *defect correction period* is set out in the Contract Data. If the *Contractor* does not correct these within this period then the *Project Manager* assesses what it will cost the *Employer* to get these corrected (by other people) and deducts that amount from the *Contractor* (see clause 45.1).

53. Disputing the Defect (clause 43)

The ECC guidance notes state that the term 'Defect' has a restricted definition and that it is intended to include unfinished or omitted parts of the *works*. We currently have an ECC Option A contract, which is in the period between Completion and the *defects date*, and as the *Employer* we are experiencing issues with the roof of a building constructed by the *Contractor*. The Works Information required a slate roof be provided for the building, with the *Contractor* responsible for the design. Following Completion, we have had dozens of tiles falling from the roof, which we view as being caused by a poor batten design within the roof structure that allows wind to blow into the void and disturb the tile fixings. However, this issue has only been discovered after take over and therefore we are seeking advice as to whether this can be a Defect for notification following Completion. The *Contractor* is contending that they are not responsible for repairing the roof or resolving the underlying cause. If the issue is classed as a Defect and the *Contractor* maintains their position then we propose to invoke our right under clause 45.1 to assess the cost of having the Defect corrected by other people and deduct this from the *Contractor*.

This certainly is a Defect. The Works Information requires that they design and install a functional roof and they have clearly failed to do so. The fact that this has not happened until after Completion (and/or take over) is irrelevant. And in any event clause 82.1 requires that until the Defect Certificate is issued (see 11.2(6) and 43.3)) the *Contractor* is required to repair damage to the *works*.

The *Supervisor* (not *Project Manager*) should immediately notify this as a Defect, if he or she has not already done so. If the *Contractor* does not correct the Defect within the *defect correction period* then the *Project Manager* can deduct the cost of getting others to repair it.

54. Quality control in the ECC (clause 43)

Does the ECC automatically or otherwise give the *Employer* a 'right to reject' the work done by the *Contractor* if it does not meet agreed requirements?

The main obligation on the *Contractor* in terms of quality arises in clause 20.1, which states the *Contractor* Provides the Works in accordance with the Works Information. A Defect in clause 11.2(5) is a part of the *works* that is not in accordance with the Works Information.

There is a *Supervisor* generally appointed by the *Employer*. The *Supervisor* and *Contractor* have a reciprocal obligation under clause 42.2 to notify each other as soon as they find a Defect up to the *defects date*.

Clause 43.1 gives the obligation to the *Contractor* to correct the Defect. Clause 43.2 states by when this is to be corrected and there are other clauses dealing with uncorrected Defects and accepting Defects.

55. Consequential costs of Defects (clause 43)

 Testing and Defects are covered under section 4 of the ECC contract but is there any provision in the contract to recover or offset the cost for damage caused by the *Contractor* or the recovery of consequential loss by the *Employer*?

 The ECC sets a period during which the Parties have dual obligations if a Defect is found. In the ECC that period ends at what the contract calls the *defects date*. Up until then the *Contractor* is required to correct any Defects that are found (clause 43.2), and the *Employer* is obliged to provide access so the *Contractor* can (clause 43.4). Neither can 'charge' the other for this, and they each cover their own costs. So the *Employer* is unable to recover consequential costs. Latent Defects found after the *defects date* will be dealt with as breaches of contract, and the dual obligation no longer applies. Damage caused by the *Contractor* when correcting a Defect will be at the *Contractor*'s risk – see fourth main bullet of clauses 80.1 and 81.1. The *Contractor* is required to insure against this (see clause 84.2). In addition, the *Contractor* indemnifies the *Employer* against this damage (clause 83.1).

56. Defects arising after the Defects Certificate is issued (clause 43.3)

 I have the job of closing out schemes that have been completed several years ago but, for various reasons, are still in an aftercare and landscaping period, although the Defects Certificate has been issued. The majority of the schemes are ECC Option C and signed by deed, which I understand has a 12 years latent Defects period. My understanding of a latent Defect is something that is not obvious, e.g. it could be a geotechnical Defect. If we suspect a Defect has arisen after the Defects Certificate has been issued, but within the 12 years period, what should we do?

 The contract says nothing about Defects after the issue of the Defects Certificate. The issue of the Defects Certificate does not relieve the *Contractor* of all of its obligations with regards to Defects – see clause 43.3. You will need to get legal advice on the *Contractor* remaining liable for breaches as you will have to show that whatever went wrong was caused by the breach of the contract by the *Contractor* and work within the Limitation Act. The contract is, quite deliberately, silent on the matter, and has no procedures to deal with it, in order to allow whatever law the contract is under to do so instead.

57. Accepting a Defect! (clause 44)

Q Our *Contractor* is proposing that a Defect is not corrected on our project and suggested that they could offer a cost saving to us. Is this permissible under the ECC and how should it be valued?

A Yes, either the *Contractor* or *Project Manager* may each propose to the other that the Works Information is changed so that a Defect does not have to be corrected (clause 44.1).

The proposal from the *Contractor* may result in an earlier Completion Date and reduction in Prices. There is no methodology as such stated in the ECC (like clause 63.1 requires for the assessment of compensation events); however, the figure needs to be sensible and agreed between the *Project Manager* and *Contractor*.

Also, when considering the monetary assessment the potential cost of not doing the work should be considered (if it is simply an omission/the *Contractor* has failed to do something). In addition, life cycle cost issues need to be considered. Under clause 44.2, once accepted the Works Information is changed accordingly – this means that the *Contractor* is no longer responsible for the Defect. If the Defect being accepted now incurs additional life cycle cost then this should also be a factor in the reduced Prices; e.g. the installation of a less efficient pump may result in a higher utility cost to run.

58. Not getting access to correct Defects (clause 45)

We are under the ECC Option C form of contract. As a *Contractor* we have been notified of a Defect although we have not yet achieved Completion. The correction of Defects under these circumstances would be a Defined Cost and therefore part of the Price for Work Done to Date. However, what stops the *Project Manager* from deliberately not granting access for the correction of a Defect, thus ensuring the *Contractor* cannot correct the Defect (and include in Price for Work Done to Date) and just use clause 45.2 to charge the *Contractor* the cost of correcting the Defect. This would appear to prejudice the *Contractor*'s position in being paid (all be it included in the calculation of the *Contractor*'s share) and just artificially creating a situation in which the *Contractor* would have to pay for the correction of all Defects.

We think you may be confusing two different matters: whether or not you get paid to carry out work, and what your obligations are to correct Defects. If, because of the definition of Completion, the Defect requires correction before Completion can be certified then the *Project Manager* cannot certify Completion until it has been corrected.

Therefore we cannot see how this scenario can happen, because it is not a Disallowed Cost until after Completion has been certified in accordance with the contract. In addition, the *Employer* is required to provide you with access to and use of the Site (clause 33.1) up until take over and if not that will be a compensation event in itself – see clause 60.1(2). You have such access and use until take over and just because the works are taken over does not mean that Completion has been achieved. Clause 45.2 only applies once Completion is certified. After that you are required to correct Defects – see clause 43.2, and under Option C the cost of that will be a Disallowed Cost.

So, after Completion, Defects are at your cost anyway, with or without clause 45.2. You need to be aware that the wording of clause 45.2 is deliberately different to that in clause 45.1. If the *Employer* cannot provide access for you to correct it the value of the deduction is not the cost of getting somebody else to correct the Defect; it is what it would have cost you to correct. This means that you are in no worse a position than if you had carried out the correction yourself.

59. Whether the *Contractor* comes back or not (clause 45)

I am trying to understand the difference between clauses 45.1 and 45.2 of the ECC. They do not make sense to me – can you help?

Yes, these outline two scenarios that relate to uncorrected Defects. Scenario 1 (clause 45.1) is when the *Contractor* was given access to correct the Defect by the *Employer* but they failed to rectify in the *defect correction period*. In this instance the monetary value is assessed as the cost incurred by the *Employer* of having the Defect corrected by other people – this will typically be a higher cost than clause 45.2. Whether the Defect is corrected by other people is up to the *Employer* but the monetary value is assessed in this manner.

Scenario 2 (clause 45.2) is when the *Contractor* was not given access by the *Employer* – perhaps this is because the building was operational and it would be impossible/unsafe to rectify the Defect. In this instance, the lesser value would be assessed – this cost was incurred by the original *Contractor* in rectifying the Defect. In theory this should be a lesser cost than clause 45.1. Again, it is then the *Employer*'s decision as to whether the Defect is actually corrected.

Both scenarios provide a sum of money for the *Employer* then to deal with the problem.

Some further thoughts and top tips

- The Works information should clearly define the quality standards.
- Ensure the *defect correction period(s)* are appropriate and thought through.
- Ensure that the *Employer* manages Defects post-Completion with its resources. Often they are not managed as the contract requires.
- If tests or inspections are undertaken by the *Supervisor* include their timescales and durations.
- Do not be afraid to notify Defects. It is what the contract requires. Following the contract is professional and provides a clear audit trail.

Chapter 5
Payment

60. Making applications for payment (clause 50.1)

Q In an ECC Option C contract, once the Completion certificate has been issued, is there a set period in which the *Contractor* can still submit applications for payment? Once the Completion certificate has been issued, does the *Contractor* have 4 weeks to submit their application and after this date if they are still working on the final application, can they still submit an application or is it then at their own cost?

A Certification and payment under the ECC does not depend on the *Contractor* making an application. Instead, the *Project Manager* assesses the amount due at the intervals set out in clause 50.1 with or without an application – see clause 50.1. If the *Contractor* has made an application, the *Project Manager* must consider it (clause 50.4), but if the *Contractor* has not, the *Project Manager* still has to make his or her own assessment. Assessment dates continue at the maximum of the assessment intervals from the *starting date* until 4 weeks after the issue of the Defects Certificate by the *Supervisor* – again see clause 50.1. So, assessments and payments are made by the *Project Manager* after Completion, to the extent that the Price for Work Done to Date has changed. And that is especially so with Option C in which you cannot make the final payment of the *Contractor*'s share until you have finalised the Price for Work Done to Date and the final total of the Prices – see clause 53.4. The last assessment and payment is made no later than 4 weeks after the Defects Certificate is issued.

61. Calculating the final account (clause 50.1)

We are engaged as *Project Manager* on an ECC Option C project in which the *Contractor* has failed to provide its final account. Completion was achieved a while ago. Do we have to provide any notice before we make a *Project Manager*'s assessment for the final account?

There is no such thing as a final account application in the ECC, unlike other standard forms of contract. In fact, the ECC does not oblige the *Contractor* to make any application for payment at all.

Instead, the obligation is on the *Project Manager* to make his or her own assessment at each assessment date (see clause 50.1). That obligation continues before and after Completion until 4 weeks after the *Supervisor* issues the Defects Certificate (see also clause 50.1). Having made his or her own assessment, the *Project Manager* is obliged to certify payment within one week of each assessment date (clause 51.1). If the *Contractor* makes an application at any time then the *Project Manager* is obliged to consider it when making his or her assessment and must give details to the *Contractor* for each assessment (clause 50.4).

So you should already be assessing what is due and certifying it. In Option C it is important you set up a system whereby the *Contractor* shows the *Project Manager* its Defined Cost each month and the *Project Manager* regularly audits this. The *Project Manager* will fail in his or her duties if he or she certifies payment based on 'say' or unsubstantiated figures – see the first bullet of the definition of Disallowed Cost in clause 11.2(25). This exercise should not be left until the end of the contract as it can lead to all sorts of problems arguing retrospectively about what is and is not Defined Cost and Disallowed Cost.

Therefore, by the time you get to completion under Option C, all of the hard work should be done and the *Project Manager* should have a good idea of what the total Defined Cost is. And until such time as the *Contractor* justifies any other costs, they are not due because they will be a Disallowed Cost until they are justified.

As to the payment of share (one way or other) the main tranche of that should have been certified by the *Project Manager* at the assessment immediately following Completion, and should be based on their forecasts of the final Price for Work Done to Date and the final total of the Prices – see clause 53.3. The final payment of the share is made as soon as the final figures for the Price for Work Done to Date and total of the Prices are known (clause 53.4).

The final payment has to be certified within 4 weeks of the *Supervisor*'s Defects Certificate (clause 11.2(6)), which in turn has to be issued by the later of the *defects date* and the end of the last *defect correction period*. So, the *Project Manager* should make his or her final assessment of the amount due by then.

62. Assessing the amount due (clause 50.2)

We have an ECC Option A contract that incorporates secondary Option X7 delay damages and X16 retention. When assessing the amount due, should the *Project Manager* deduct retention before or after applying delay damages?

The amount due at each assessment date is detailed in the three bullets of clause 50.2. It states it is the Price for Work Done to Date, plus other amounts to be paid to the *Contractor*, less other amounts to be paid by the *Contractor*. Clause X16.1 makes it clear you deduct retention from the Price for Work Done to Date less the retention-free amount. It is not deducted from the other amounts in the other two bullets of clause 50.2. You therefore calculate the Price for Work Done to Date, then deduct the retention. After that, the delay damages are deducted under the third bullet of clause 50.2. The *Project Manager* therefore deducts retention before applying the delay damages.

63. Retaining money due to non-compliant programmes (clause 50.3)

On an ECC Option C contract we have a query related to the Accepted Programme. At the outset of the contract we had a programme identified in Contract Data part two. This was further developed by the *Contractor* to provide more detailed information, i.e. it went to a lower level to identify the deliverables associated with the associated tasks. This programme was then used by all Parties as the managing tool and was updated and issued on a monthly basis as required under clauses 32.1 and 32.2.

The project has now been going for some 4½ years and in this period the programme has only been formally rejected twice. There has never been a formal acceptance of any programme. While the programme issued under clause 32 and the one in the Contract Data part two has been used by both Parties, they have never strictly complied with the requirements of clause 31.2.

The *Employer* is now proposing to retain 25% of the Price for Work Done to Date in accordance with clause 50.3. While we appreciate this should only be retained in relation to the first programme, can this still be applied, as the programme is not in accordance with clause 30.1 and has never been formally accepted?

The simple answer is no, the *Employer* cannot do that. Clause 50.3, which deals with the possible retaining of 25%, is very clear. It starts with the words 'If no programme is identified in the Contract Data...'. As there is a programme identified in the Contract Data, clause 50.3 simply does not apply at all.

It is quite surprising that the *Employer* is even suggesting that they should do this. For the last 4½ years the *Contractor* has done what the contract requires of them, in that they have produced and submitted programmes for acceptance at the interval stated in the contract. In contrast, during that time the *Project Manager* (who works on behalf of the *Employer*) has failed almost completely to do what he or she is supposed to do under the contract. The *Project Manager* is supposed either to accept or reject each programme within 2 weeks of issue (clause 31.3). Instead he or she has, for the most part, ignored the submissions, which are breaches of contract on his or her part that have resulted in a whole series of potential compensation events under clause 60.1(6). To suggest that the *Contractor* should now be punished for some theoretical wrong at this late stage does not seem to be sustainable.

In addition, that failure now puts the *Employer* in a very weak position when it comes to disputes as to time. To suggest later that the programmes issued by the *Contractor* for acceptance were somehow 'wrong' will be very difficult, given that, with two exceptions, the *Project Manager* has never rejected them, as the contract requires he or she does if they are wrong, and then those supposedly 'wrong' programmes were used by both sides to manage the *works*.

64. No compliant programme is ever submitted (clause 50.3)

Clause 50.3 of the ECC states that 'If no programme is identified in the Contract Data, one quarter of the Price for Work Done to Date is retained in assessments of the amount due until the *Contractor* has submitted a first programme to the *Project Manager* for acceptance showing the information which this contract requires.'

My question is, what happens to the retained money should the *Contractor* fail to submit a compliant programme (or indeed, a programme of any kind) throughout the contract period? In anticipation of the possible reply that a *Project Manager* should not allow the contract to continue without an Accepted Programme, in our case the *Employer* is unlikely to want termination on the grounds that the *Contractor* has not provided a programme – particularly when works are seen to be progressing in an apparently organised manner.

Without an Accepted Programme the contract cannot be properly managed. A number of compensation events rely on an Accepted Programme – see clauses 60.1(2), (3) and (5). And without it, how can you make clause 63.3 work in respect of moving the Completion Date?

The lack of a programme is indicative of one of two things. Either the *Contractor* is disorganised, despite appearances to the contrary, and he or she has no way of knowing whether or not it is going to finish by the Completion Date. Or the *Contractor* has a programme that it does not want to show you so that it can do things the 'old fashioned' way, i.e. make one up afterwards once it knows where the delays have occurred, to ensure the critical path runs through all of those delays that you caused, and none that the *Contractor* caused. Neither is good news for your project and both should be avoided at all costs.

On the other hand, the *Project Manager* should not use the list in clause 31.2 as a 'rule book'. As long as he or she is satisfied that the programme is achievable and has all the information he or she needs to administer the contract on behalf of the *Employer* (such as where the critical path is and what resources have been assumed) then he or she should accept it, and release the 25% being retained. Without an Accepted Programme, the assessment of all compensation events is taken out of the hands of the *Contractor* altogether. The *Project Manager* is required not to accept any compensation event quotations and make his or her own assessment of the compensation event (see third and fourth bullets of clause 64.1). And that assessment must be made using the *Project Manager*'s assessment of the programme for the remaining works – clause 64.2.

As for when the 25% is released, the contract is silent, because it did not envisage that any *Contractor* would allow it to be applied for any length of time, given that cash flow is the life blood of any *Contractor*'s business. Because there is no date for release other than the production of a programme, there is an argument that says that the *Employer* keeps it, although it may not get a sympathetic ear from some adjudicators if the *Contractor* were to dispute that.

We would strongly suggest that the *Project Manager* actively manages this situation. Without a programme there is a real risk that Completion will be delayed. The *Project Manager* should notify an early warning to the *Contractor*, ideally attending a risk reduction meeting soonest with senior members of the *Contractor*'s organisation in order to sort this problem out.

65. Retaining monies when programme not accepted (clause 50.3)

On an ECC contract we have an initial programme; however, we have since had compensation events for which we believe we have not received adequate particulars covering additional cost or programme alterations. The latest revised programme purely shows a period of time risk allowance having been used up and the effects of the compensation events are still unknown. We consider this programme is not acceptable. Can the *Project Manager* therefore deduct 25% of the monies owed to the *Contractor* for this, or does this provision apply to the first programme only?

If the initial programme was supplied and referred to in the Contract Data part two under the heading 'The programme identified in the Contract Data is...', then you cannot apply the 25% retention. This is because clause 50.3 makes it clear that it only applies if no programme is identified in the contract data. However, if

- there was a programme issued with the tender that was not referred to in the Contract Data, and
- the programme you now have is the first one issued for acceptance since the contract came into existence, and
- the programme does not show all of the matters listed in clause 31.2,

then you can apply the 25% retention.

There are though, two other things you need to be aware of. Firstly, this retention does not depend on the *Project Manager* accepting the programme. As long as it has been issued and shows all the information it should, then the retention is not taken even if the *Project Manager* thinks it is wrong for some other reason. Secondly, this is only a retention, not a deduction, and eventually it will have to be paid back.

66. Pay less notices and *delay damages* in the ECSC (clause 50.3)

Under the ECSC, can *delay damages* be deducted without any need for a pay less notice to be issued?

We cannot give legal advice so we are limited in what we can say on this matter. This will come down to a legal interpretation of legislation. Looking at clause 50.3, the deduction of *delay damages* is part of the amount due (see the third bullet). However, this contract works on the trigger for payment being the *Contractor*'s application – see clauses 50.1 and 51.1. And the contract requires that you must notify the *Contractor* before you pay if you are not going to pay the amount in their application (clause 50.4). We assume that you did not do that, which is a problem, because now you are in breach of contract. However, it could be argued that you did not need to provide a pay less notice for this, because it was not part of the amount due and therefore you were not deducting it, you were just not paying it. That would be an argument that may be looked at sceptically given that you have not done what clause 50.4 requires.

We would therefore strongly recommend that you always provide a notice of what you are going to pay if it is not what the *Contractor* has applied for, with details of what deductions you are making and why.

67. Error in the ECSC Price List (clause 50.3)

I have a query regarding the ECSC. If a tenderer makes an arithmetical error in the Price List, should this error be corrected when the tenders are opened and the revised figure inserted in the *Contractor*'s Offer? Or can the tendered sum not be changed?

It is entirely up to you how you deal with it at tender stage. You have no contract at tender stage so there are no rules that are relevant within the contract (the same as other contracts). It will also depend on what your client's policy is on the matter, and what the Instructions to Tenderers said you would do in the circumstances. Some clients adjust the total, other clients ask the tenderer if they wish to stand by their total, and if so what adjustments they want to make in the prices in the Price List. Some clients would leave the figures as is but use the correct total to assess the tenders. None is wrong or right. As to what happens if you accept the tender with this error in it, the contract is clear. The *Contractor* is paid the Price for Work Done to Date – see clause 50.3. The definition of Price for Work Done to Date is in clause 11.2(9), and you will see it is based solely on the figures in the Price List. So the *Contractor* will get paid the correct total, not the incorrect figure they added it up to be at tender stage.

68. Difference in payment periods between ECC and ECSC (clause 51)

In the ECC clause 51.2 states that the *Employer* must pay the *Contractor* within 3 weeks of the assessment date. The ECSC clause 51.1 states that the *Employer* must pay the *Contractor* within 7 weeks of the assessment *day*. Why are these so different?

We disagree with your opinion that the ECSC provides for a 7-week payment provision. It provides for 3 weeks after the next *assessment day*. Clause 50.1 states that the *Contractor* has to make an application before each *assessment day*. So, as long as the *Contractor* gets their assessment of the amount due in to the *Employer* by the day before the *assessment day*, they will be paid 3 weeks later. However, if they miss that deadline and do not get it in until the day after then they have to wait until the following month. So as long as the *Contractor* does what clause 50.1 of the contract requires, they will be paid in 3 weeks. If they do not, they have to wait another month. So there is no difference, as long as the *Contractor* follows the requirements of the contract.

69. Can the *Contractor* owe money to the *Employer*? (clause 51.2)

In the ECC, should the *Contractor* be required to pay damages from one of the secondary Option clauses or amounts for uncorrected Defects, and is the *Project Manager* required to certify as per the requirements of the core clauses in section 5 by reducing the amount due by the relevant amount, and therefore creating a situation in which the *Contractor* would owe the *Employer* that amount due?

To add some context to this, we have a situation in which damages and uncorrected Defects may create a situation that the *Contractor* has to pay the *Employer*. The *Project Manager* has decided that he or she can just demand payment almost immediately and without any form of certification; we state that the provisions and payment times are dictated by the core clauses in section 5.

The only way that payment becomes due from one Party to the other is through the operation of the valuation and certification process set out in section 5 ('Payment') section of the contract. The *Project Manager* is required to issue a certificate for the amount due in accordance with clause 50. The timing is set out in clause 50.1 and runs all the way until 4 weeks after the issue of the Defects Certificate. The amount due is set out in clause 50.2. The sums you are referring to would fall under the third bullet of clause 50.2 'less amounts to be paid by or retained from the *Contractor*'. The *Project Manager* then has to certify in accordance with clause 51.1.

If that certificate reduces the amount due to from that previously paid by the *Employer* then clause 51.1 requires the *Contractor* to make the payment within the time scales set out in clause 51.2. If you do not pay by that date interest becomes due to the *Employer* at the rate in the contract, just as it would become due to you if the *Employer* does not pay on time.

When deductions can be made depends on what they are being made for. The earliest the *defect correction period* can run from is Completion (see clause 43.1), and so there can be no deductions for your failure to correct Defects until after Completion. The same applies with delay damages. If the *Project Manager* wants to recover these from you then he or she has to assess them and include them in the next assessment he or she is due to make under clause 50.1. That will usually be monthly. The *Project Manager* then follows the processes in clauses 50 and 51, and you pay in accordance with clauses 51.1 and 51.2. By the way, if you do not intend to pay you must issue a pay less notice in accordance with Y2.3 (or whatever other part of the contract complies with the HGCRA). The requirement for such a notice falls on whichever Party is not intending to pay, not just the *Employer*.

70. Payment in the ECSC (clause 51.2)

We are currently working under an ECSC. The *Employer* has stated that under clause 51.1, they can take up to a further 3 weeks to pay after the next *assessment day*, which would be say 4 weeks later. This would be a total of 7 weeks from the original *assessment day*. Our normal terms under the ECC, for example, are that the *Project Manager* certifies within a week of the assessment date, and that the certified payment is actually made within 3 weeks of the assessment date (clause 51.2). The ECSC guidance notes state that payment relates to an *assessment day*, which we always assumed was the original date, but did not think it would be the next one. Can you advise please?

It seems that the *Employer* has misunderstood the contract. The ECSC uses the principal of an *assessment day*, which is set out in the Contract Data – see page 2. It could be 'the last day of the month' or the '28th of the month', or 'the last Friday of each month', or any of those sorts of descriptions. You will need to check that. You are then required to make an application for payment in each period (clause 50.2) before each *assessment day* (see clause 50.1). You will then be paid 3 weeks after the next *assessment day* following receipt by the *Employer* of your application (clause 51.1).

Let us say your assessment day is the 28th of each month and we are looking at April. If the *Employer* receives your application by the 27th April then the next *assessment day* is the 28th April and you will therefore be paid 3 weeks after the 28th April. If the *Employer* does not receive it until the 29th April, then the next assessment day is 29th May and you will be paid 3 weeks after the 29th May. That means that as long as the *Employer* receives your application before the *assessment day*, you will be paid within 3 weeks of that *assessment day*. If you miss that *assessment day* (even by a day) you have to wait until the following month.

71. Can we add corporation tax into the Fee? (clause 52.1)

 Under ECC Option E can corporation tax be added into the Fee provision?

 The ECC does not work like that. The contract defines what you will be paid for as Defined Cost (clause 11.2(24)) and then says any other cost you incur or any other monies you want to recover have to be included in your Fee (clause 52.1). Corporation tax is clearly not included within the SCC and is therefore not paid as a part of Defined Cost.

It is entirely up to you what Fee percentages you insert in the contract and what allowances you want to include within your Fee percentages and at what level. You can add whatever you want to the Fee percentages and once they are set in the contract that is the end of the matter and the *Employer* cannot audit it in any way.

All the ECC guidance notes point out is what is not included within Defined Cost and which you therefore need to think about when assessing what Fee to quote. They are not saying that you should or should not allow for any of the listed items in your Fee, because that is a commercial decision only you can make. However, if you do not allow for them you will not get paid for them anywhere else.

72. Passing on discounts (clause 52.1)

We are the *Contractor* on an ECC Option B contract. A compensation event has arisen that includes some Subcontractor works (the Subcontractor is already on Site). We have previously agreed a 2.5% main *Contractor*'s discount with the Subcontractor for prompt payment. Are we contractually required to pass this discount on to the *Employer*?

The simple answer is yes, see clause 52.1 – 'with deductions for all discounts rebates and taxes which can be recovered'. However, when assessing compensation events in Option B it is, in any event, irrelevant what you have paid (or will pay) the Subcontractor. Your calculation has to be based on the Defined Cost to the Subcontractor for carrying out the *works* calculated based on the SSCC in your (i.e. main) contract – see the definition of Defined Cost in clause 11.2(22).

Some further thoughts and top tips

- Encourage the *Contractor* to make sure that they do not fall foul of clause 50.3 by submitting a first programme showing the information that the contract requires.
- The *Project Manager* should be aware that he or she has to determine the amount due at each assessment date, with or without the assistance of any application for payment by the *Contractor*.
- When interest is due, this does not have to be asked for by the receiving Party, it should be included in the next assessment.

# Chapter 6
Compensation events

73. Omitting work under ECC Option A (clause 60.1(1))

Q If we omit work from the Works Information under ECC Option A does this result in a negative compensation event?

A Any change to the Works Information (under clause 14.3) is a compensation event (with the exception of the two bullets under clause 60.1(1)). If you omit work then the forecast cost saving will be assessed in accordance with clause 63.1.

74. Inconsistencies in the Works Information (clause 60.1(1))

Q We have just entered into an ECC Option C contract and immediately encountered a problem with the *Project Manager*. The Works Information prepared by the *Employer* (part 2) requires that the heating system is on two zones whereas our price and our Works Information (part 1) states one zone. From our point of view what we have offered has been accepted and therefore that is what we believe to be the basis of the contract. The *Project Manager* disagrees. Please can you confirm our understanding?

A Unfortunately your understanding does not accord with the contract. Essentially what we have is an inconsistency between the Works Information prepared by the *Employer* and the Works Information prepared by the *Contractor*. This needs to be notified under clause 17.1, the *Project Manager* will then decide which is preferred. A change to the Works Information is then instructed under clause 14.3. The answer is in clause 60.1(1) second bullet – this states that a change to the Works Information prepared by the *Contractor* (your part 2), which is required in order to comply with Works Information prepared by the *Employer* (your part 1) is not a compensation event. So, the *Project Manager* is right, you will need to install two zones and you will not be entitled to a change in the Prices or Completion Date. The cost, however, will be a shared risk under Option C, dependent on the share ranges.

75. Why is accepting a Defect not a compensation event? (clause 60.1(1))

I do not understand the first bullet under clause 60.1(1) under the ECC – why is acceptance of Defect not a compensation event?

A Defect (as defined in clause 11.2(5)) may be accepted under clauses 44.1 and 44.2. It is not a compensation event as it does not follow the assessment methodology defined under clause 63.1 and other clauses.

A quotation for reduced Prices or an earlier Completion Date or both is submitted to the *Project Manager* for acceptance. This is effectively a commercial offer and, from the *Project Manager*'s viewpoint, needs to address future maintenance costs. For this reason it is not subject to the same assessment methodology as a compensation event.

76. Are compensation events claims? (clause 60.1)

This may seem like a basic question but, being new to the ECC, what are compensation events? Are they claims?

It is a good question – compensation events are a defined list of events and should they occur, the *Contractor* can be compensated potentially in terms of both time and cost. Within any contract we need to define the limit of risk. The ECC defines 19 (in the core clauses) events that are defined as compensation events. Additional compensation events are also found in the main and secondary Options.

Compensation events broadly fall into three main categories – we have changed our mind, we have not done something we said we would, and finally a risk has occurred over and above a reasonable risk limit of a competent contractor.

The whole premise of compensation events is that the *Contractor* should be no better or worse off after the compensation event. You will see within the assessment clauses that we assess both time and cost. The other key difference with most traditional contracts is that compensation events are assessed and what is termed 'implemented' within defined timescales.

77. Stopping the work (clause 60.1(4))

We are working on a petrochemical site and are under Option A of the ECC. The *Project Manager*, due to other projects on the Site, has on a number of occasions told us to stop work and on other occasions prevented us from doing work. There was nothing mentioned in the Works Information about this – we have incurred cost as a result. Is this a compensation event?

Yes, it is important to note that you need to obey these instructions (clause 27.3), which it sounds like you have. The *Project Manager* has the authority to give such an instruction but you are right – this is a compensation event under clause 60.1(4). This should be notified by yourselves under clause 61.3 (if not already done by the *Project Manager*). This will start the compensation event process and enable the disruption to be assessed.

78. Dealing with Others (clause 60.1(5))

We are delivering a complex engineering project under Option C of the ECC. The client will be involved in reviewing our design as a *Contractor*. The work also encroaches on a railway line and so will involve the rail operator. We also have potential archaeological issues and utility connections. How do we define the interaction of all these key players?

The answer to this is that – it is all in the programme! The programme is at the 'heart' of the ECC. Clause 31.2 requires the *Contractor* to chart (fourth bullet) the order and timing of the *Employer* and Others. The Others are defined (clause 11.2(10)) and basically cover organisations that are not employed by the *Contractor*, i.e. they are the *Employer*'s responsibility.

Failure of the *Employer* or Others to work within the times shown on the Accepted Programme, failure to work within the conditions stated in the Works Information or the carrying out of additional work would all fall under the realm of a compensation event – clause 60.1(5). It is important that the programme is managed collaboratively and that the interaction of all these organisations is clearly defined. A suggestion would be to meet with the *Project Manager* and discuss the programme before submission – this will ensure it is accurate and agreed.

79. Late decisions (clause 60.1(6))

I am a *Project Manager* on an ECC Option A contract. I am aware of the need to make decisions within defined timescales but the client I am working for is notorious for not delegating and taking too long to make decisions. I am delivering a Board presentation to the *Employer* team and want to highlight the key risks – please can you advise what the impact of late decisions would be?

You are right that the ECC requires effective decision-making – this is fundamental to any project management approach. Ultimately, failure to respond to a communication is a compensation event (clause 60.1(6)). Let us look at three examples, failure to respond to a design submission will stop the job – the *Contractor* does not proceed with the relevant work (clause 21.2); failure to respond to proposed Subcontractors will prevent the orders being placed – again, having a material impact. However, failure to respond to the programme (clause 31.3) does not 'stop the project'. The *Contractor* would still be obliged to continue. In all instances, however, the *Project Manager* should respond within the period required by the contract. This is a key message to get across to your Board. The ECC assumes that full authority is placed with the *Project Manager*.

80. Objects of interest (clause 60.1(7))

We have a situation on our project in which we found an object of historical interest yet it was in a corner of the site that did not affect the *Contractor*. So, in principle, it is a compensation event but there was no time or cost impact. How do we close this off? As *Project Manager* I do not want to leave things open ended for someone else to deal with later or new people having different interpretations.

OK, what you need to do is notify the compensation event under clause 61.1 and instruct the *Contractor* to submit a quotation. This gives the *Contractor* opportunity to assess the time and cost impact. Subject to this coming back as no time/cost impact this will be implemented as such (clause 65.1 first bullet) and this is not then revised (clause 65.2). You are right to close things off as instructions from the *Project Manager* are not subject to the clause 63.1 time bar.

81. Not accepting the design (clause 60.1(9))

I am trying to understand clause 60.1(9) and think of an instance when this could apply – please can you advise?

OK, let us say that we have asked a *Contractor* to build a new football stadium. In the Works Information we have asked him to design all of the *works* (clause 21.1 refers). We then require him to submit the steelwork design (as per clause 21.2). In the Works Information we do not impose any constraints or mention the potential for future expansion to the football stadium. When the steelwork drawing comes through we then realise, based on the structure, it would prevent future expansion. Our only two reasons for not accepting (mentioned under clause 2.1) are that it does not comply with the Works Information or the applicable law – we therefore cannot 'not accept' based on those two reasons, so we have to 'not accept' for 'a reason not stated in this contract'. We can do so but then, as is fair, this would give rise to a compensation event.

82. Tests and inspections that the Works Information requires (clause 60.1(10))

I am a frustrated *Supervisor* on an Option C ECC project! The *Contractor* is hardly doing any tests and inspections – only the bare minimum required by law and I suspect that there may be Defects. What can I do? Can I request to search for suspected Defects?

This all comes back to drafting the Works Information professionally at the outset – it sounds like you were not given the opportunity to input into this. Clause 40.1 states that the test and inspection clauses only apply to those required by the Works Information or applicable law.

You can get the *Project Manager* to change the Works Information (clause 14.3) but this will be a compensation event (clause 60.1(1)). An alternative approach would be to get the *Project Manager* to notify an early warning and get involved in discussions about the potential impact of such a change.

If you do suspect Defects then you have the power as *Supervisor* to instruct a search. However, if none is found it is a compensation event.

The best approach here is to deal with things proactively and perhaps change the Works Information to reflect the tests and inspections that you feel are appropriate (this needs to be by means of the *Project Manager*). Clause 40.3 then requires the *Contractor* to notify these before they are undertaken and the result afterwards.

83. Unnecessary time for tests and inspections (clause 60.1(11))

We are a *Contractor* on a sports centre project and the *Supervisor* is taking what we believe to be an unnecessary amount of time to do his tests and inspections. Where do we stand with this?

Ideally the Works Information would have stated the tests and inspections that were required (clause 40.1) and also how long they would take. The tests and inspections normally fall into three categories – those that the *Contractor* is required to undertake, those undertaken by the *Supervisor* and those that the *Supervisor* wishes to witness. If a timescale is not stated in the Works Information then it may come down to interpretation/common practice. If you feel that this is a compensation event then we would notify as such (clause 61.3 requires this to be undertaken within 8 weeks of becoming aware). As it seems to be a repeat problem/trend, we would also notify an early warning and try to resolve proactively.

84. Interpreting physical conditions (clause 60.1(12))

Q We have a *Contractor* on Site under ECC Option A. The work involves driving along a quay, down a ramp and onto the beach – they are undertaking jetty repairs to one of our docks. They have recently notified a compensation event under clause 60.1(12) for a 'Max Headroom' goal post steel frame, which prevents certain sized Equipment using the ramp. We did not mention this in the Site Information but they were allowed to undertake a Site visit as part of the tender. Please can you advise where we stand with this?

A The two clauses that need to be in conjunction with one another are clause 60.1(12) and clause 60.2. Clause 60.2 explains that 'in judging a physical condition' the *Contractor* is assumed to have taken into account information obtainable from a visual inspection. So just what is contained in the Site Information is not the whole picture. If it could have been seen from a visual inspection (which we are sure this could) then it would not be a compensation event.

85. Underground mine workings? (clause 60.1(12))

We are a local authority about to use an Option A ECC contract to procure the installation of a flood wall in a coastal area of Scotland. The area is notorious for underground mine workings but we do not have accurate records as they were not kept in the past. We do not want, on discovery of underground mine workings, this to be a source of debate nor do we want the tenderers to price for this and for it still to be debated as a compensation event. Do you have any suggestions? We want to be a fair procurer and not force undue risk on to the tenderers.

Two options exist – you could either introduce a Z clause to clarify the definition or add an additional *Employer*'s risk in Contract Data part one. The additional *Employer*'s risk is perhaps best in this situation. This could be worded as 'Discovery of underground mine workings' (or perhaps clearer wording). If this occurs this would then be a compensation event under clause 60.1(14). This will ensure that the *Contractors* do not price this risk in the tender – the only disadvantage is that if the risk occurs it will be a compensation event. The main principle is that it is clear and, as per your objective, fair.

86. What are physical conditions? (clause 60.1(12))

In an ECC contract, what is the accepted definition of 'physical conditions' in clause 60.1(12)? The ICE conditions of contract used to refer to 'artificial obstruction', I think it was, which would cover uncharted utilities. Are there circumstances in which physical conditions under ECC could cover uncharted services?

A glance at a dictionary says that physical in the sense used here means 'of material things or nature'. An artificial obstruction is a 'material thing' in the same way as a natural obstruction or the ground itself is. It is not clear why the now defunct ICE conditions used both 'physical conditions' as well as 'artificial obstructions' in clause 10 because the latter is already covered by the former. Maybe it was just a matter of emphasis or belt and braces. Either way it is not needed nor used in the ECC, because physical conditions would include artificial obstructions. A pipe or cable, whether charted or uncharted, is a physical condition and we have never heard anybody seriously suggest otherwise.

87. Weather risk (clause 60.1(13))

We are a large contractor and want to understand the risk as defined under clause 60.1(13) of the ECC. Please can you clarify?

We have a scientific assessment of weather in the ECC. First of all the weather measurements are those detailed in the Contract Data part one so these need to be reviewed for each tender (note that wind is not included by default). Only the weather measurements detailed in Contract Data part one are relevant in assessing a compensation event. We look at each month in isolation so bad weather spanning 2 months would be reviewed separately as two compensation events. It is also only relevant if before the Completion Date – so if you are late then all weather is the *Contractor*'s risk.

Depending on who is providing the weather data (which will be defined in Contract Data part one) it may be calculated in different ways. If it is the UK Meteorological Office then a long-term 30-year period is typically assessed and the worst three measurements averaged to give the weather shown to occur 'on average less frequently than once in ten years'. This is also known as the 1 in 10 return value – details of ECC weather data and how it is calculated are found on their website. The good news is that it can be scientifically assessed both at tender and for compensation events. The background risk, however, is not insignificant so needs to be considered carefully in your tender programme and price.

88. Take over (clause 60.1(15))

We are delivering a server centre and need to get into some of the rooms to store kit that has arrived early. This is under ECC Option C. Are we allowed to start using parts of the building before the *Contractor* has finished?

Yes, clause 35.2 allows the *Employer* to use part of the *works* before the Completion Date – so you have the authority. However, this will be a compensation event (clause 60.1(15)). As soon as the *Employer* begins using then take over is said to have occurred – this is then certified by the *Project Manager* within one week.

Also, from take over loss of or wear or damage then becomes an *Employer*'s risk (clause 80.1).

89. Assumptions made by the *Project Manager* (clause 60.1(17))

Q We have some complex compensation events on our ECC Option C project. As *Project Manager* I do not want the *Contractor*'s assessments to be full of risk – how can I deal with this?

A At a macro level *Employer*'s risks (Contract Data part one) change the risk profile of the contract as a whole. On a compensation-by-compensation event basis we can add assumptions as a *Project Manager* under clause 61.6. The original compensation event is not then revised if an assumption is wrong (clause 65.2). This becomes a separate negative or positive compensation event to correct the assumptions (clause 60.1(17)).

This an excellent process of getting better value from compensation events, fair risk allocation and enabling complex compensation events to be assessed.

90. Prevention? (clause 60.1(19))

 Does Subcontractor insolvency come under clause 60.1(19) of the ECC?

 The definition of clauses 19.1 and 60.1(19) needs to be carefully considered. It has to prevent the *Contractor* completing the *works* or completing the *works* by the date shown on the Accepted Programme and neither Party could predict or prevent. The wording also defines that it would have been unreasonable for the *Contractor* to have allowed for it.

Subcontractor insolvency may not fall into this category – it happens relatively frequently unfortunately. This clause is designed not to pre-empt a defined list of issues but rather describe the scenario. In order for a compensation event to be successful under clause 60.1(19) it would need to pass all of the tests contained within the clause.

91. Poor Site Information (clause 60.2)

What happens if the *Employer*'s Site Information is ambiguous or inconsistent?

This is produced by the *Employer* – they effectively verify its accuracy. This would first be notified under clause 17.1 as an ambiguity or inconsistency. If it is found to be ambiguous (vague) or inconsistent (contradictory) then it is found most favourably to the *Contractor* (clause 60.2).

92. Mistakes in a Bill of Quantities (BoQ) (clause 60.6)

We have an ECC project under Option B. There is a requirement in the Works Information for video surveys. There is not a BoQ item for video surveys. Are we entitled to a compensation event by way of the following process?

1. Inform the *Project Manager* of an ambiguity or inconsistency in or between the documents in accordance with clause 17.1.
2. The *Project Manager* must correct mistakes in the BoQ under clause 60.6 and such corrections are compensation events.
3. Clause 60.7 says that when assessing such a compensation event, the *Contractor* is assumed to have taken the BoQ as correct.
4. Therefore, we are entitled to a compensation event for the cost of the video surveys plus the Fee.

Note: all of this has just arisen and there has been no correspondence to date with the *Project Manager* on the issue.

Clause 60.6 only applies to certain 'mistakes', i.e. when they are departures from the rules and divisions set out in the *method of measurement*. The answer to your question therefore depends entirely on what the *method of measurement* says about such videos. If they are included, either explicitly or implicitly, as part of the item coverage, then there is no mistake. If they are not covered, or are mentioned as being required to be measured separately, then there is a mistake.

Clause 60.7 only applies to how the compensation event is valued, not whether or not it is a compensation event in the first place. So if it is a compensation event under clause 60.6 then, under clause 60.7, it is valued assuming that you did not allow for it in your Prices and programme.

93. Error in the Bill of Quantities caused by the *Contractor* (clause 60.6)

In a project using ECC Option B the *bill of quantities* was issued to tenderers to price. The successful *Contractor* returned the priced *bill of quantities* (which of course became the Bill of Quantities) but in a different format so that it aligned with its subcontract packages. In re-arranging the *bill of quantities* items the *Contractor* missed off some measurable *works*, according to the *method of measurement*. The Works Information has not changed but is it correct to assume this becomes a compensation event under clause 60.6, even though it is an error originally made by the *Contractor*?

You are correct; this would be a compensation event under clause 60.6. Furthermore, when assessing this compensation event, the *Contractor* is assumed to have taken the Bill of Quantities as correct (clause 60.7). In the circumstances, the changes made by the *Contractor* to the Bill of Quantities were, in effect, a qualification to its tender. Your client should have treated it as such and checked it very carefully before accepting the tender. Once accepted, the risks of such errors pass to the *Employer*.

94. The difference between clauses 61.1 and 61.2 (clauses 61.1 and 61.2)

What is the difference between a compensation event under clause 61.1 and clause 61.2?

Clause 61.1 is a clear instruction to put the work 'into effect' (last sentence). This is a clear green light to undertake whatever has been requested. An instruction to provide a quotation will also be provided at the same time and this will follow the compensation event assessment process.

Clause 61.2 is not a compensation event – it is an instruction to provide a quotation. This last sentence states that this is 'not put into effect'. We are effectively requesting a quotation and then we shall decide in due course.

An obvious comment but these need to be clearly communicated (citing the clause) in order to avoid confusion.

Compensation events

95. The time bar under clause 61.3 (clause 61.3)

As a *Contractor* under an ECC Option B contract we have notified the *Project Manager* of poor ground conditions under clause 16.1. Should he or she notify a compensation event or do we?

Early warnings and compensation events are completely separate procedures. The easiest way to think of these is that there is no time nor money in early warnings – only risk reduction. If, as a *Contractor*, you wish to be compensated with time and/or money then the only way to do this is through a compensation event. Clause 61.3 clarifies that, if you do not notify within 8 weeks of becoming aware, you will lose entitlement. This is referred to as the time bar. The exception to the time bar is if the event arose from the *Project Manager* or *Supervisor* giving an instruction, issuing a certificate, changing an earlier decision or correcting and assumption.

The clue to which of the 19 core compensation events may be time barred is the wording. For example, clause 60.1(1) would not be time barred as it would arise from an instruction from the *Project Manager*. However, clauses 60.1(2) and 60.1(12) would be and so on.

96. *Project Manager's* failure to respond (clause 61.4)

We have a situation in which the *Project Manager* is just not responding to our notifications of compensation events under Option A of the ECC. As a *Contractor* we need to get the financial and programme impact of compensation events agreed. What can we do to progress these?

Clause 61.4 provides one week for the *Project Manager* to respond to a compensation event (or longer time if agreed with the *Contractor*). Ultimately, if the *Project Manager* does not respond, the *Contractor* may (which means it is discretionary) notify the *Project Manager* of his or her failure. If the *Project Manager* fails to respond within a further 2 weeks his or her failure is treated as an instruction to provide a quotation.

It is always best to agree collaboratively if at all possible – perhaps consider notifying an early warning and bring the backlog to a head?

97. The link between early warnings and compensation events (clause 61.5)

Our *Contractor* failed to notify an early warning for a matter that was ultimately a compensation event. While undertaking refurbishment work in our office block they discovered a damp patch in a number of ceiling tiles. Without an early warning they took it upon themselves to replace the whole ceiling. However, we are frustrated as the ceiling was going to be taken out by another contractor in our employment and a painted finish applied to the services – effectively doing away with the suspended ceiling. Where do we stand with this – do we have to pay?

There are a number of clauses that are linked here (clauses 16.1, 61.5 and 63.5). Clause 16.1 obliges the *Contractor* to notify an early warning. Clause 61.5 states that, on notification of a compensation event, if you believe that an early warning should have been given then you notify this decision. This is then linked to clause 63.5, which states that, if notified under clause 61.5 (so this becomes a condition precedent), it is assessed as if the early warning had been given. So in your case you would assess the compensation event as nothing as your other contractor would have dealt with it. You have not stated which main Option you are using but if you are using C, D, E or F then the excess cost will also be disallowed, e.g. clause 11.2(25).

98. Compensation events after Completion (clause 61.7)

Our *Contractor* has successfully completed a new school for us. We certified Completion 3 months ago and the *defects date* is 52 weeks after Completion. Are we able to get them back to do additional work as a compensation event?

In short, yes! Clause 61.7 allows you to instruct compensation events up to the *defects date*. They must obey any such instruction under clause 27.3. The only down side is that the cost will be more expensive than when they were on Site. The clause 63.1 forecast of cost will include any mobilisation costs, for example.

99. Additional work after Completion (clause 61.7)

I am a *Project Manager* on an Option B ECC project. The project achieved Completion 3 months ago for which I issued the Completion certificate. The *Employer* now wants some additional work undertaken but I cannot find a clause that allows me to instruct this?

The ECC provides provision for this, so no problem. Under clause 61.7 a compensation event can be issued at any point up until the *defects date*. The only disadvantage of notifying compensation events after Completion is that the *defects date* will not be extended and they will be more costly – clause 63.1 explains that they will be on the basis of a forecast cost. Inevitably the cost to the *Contractor* will typically be greater once they have left Site.

100. Obtaining different quotes for the same compensation event (clause 62.1)

 We have a compensation event on an Option D ECC project. Are we able to get different quotations for different methods of dealing with a particular problem?

 Yes, clause 62.1 enables the *Project Manager* to obtain different quotations for the same compensation event. The wording used is 'After discussing with the *Contractor* different ways of dealing with the compensation event...', so it requires the *Contractor* and *Project Manager* to work closely and develop sensible/practical approaches.

101. Updating for remaining work (clause 62.2)

I am a bit concerned on our ECC Option A project. Clause 63.3 states that we use the latest Accepted Programme and then assess the impact of the compensation event on planned Completion. However, our problem is that the *Project Manager* has not accepted the programme for 6 months. Surely this cannot be right as we are looking at the impact on an out-of-date programme?

The ECC clauses cannot be read in isolation and this instance is a classic example. Clause 62.2 needs to be read in conjunction with clause 63.3. Clause 63.3 requires the latest Accepted Programme to be used, for this to be then updated for 'remaining work' (clause 62.2) and then the impact assessed on planned Completion (back to clause 63.3). Both clauses are used by the *Contractor* to assess a compensation event. The same methodology is used if it is the *Project Manager* that assesses – clause 64.2 requires the programme to be updated for 'remaining work'.

102. Proposed changes to which programme? (clause 62.2)

We are the *Employer* using the ECC Option C. A *Project Manager*'s instruction has been issued changing the Works Information. The *Project Manager* has also notified the *Contractor* of the compensation event and instructed the *Contractor* to submit quotations at the same time, as in clause 61.1. The *Contractor* submitted a quotation for the compensation event within 3 weeks. The *Project Manager* has instructed the *Contractor* to submit a revised quotation (clause 62.4), which the *Contractor* has also submitted within 3 weeks. The *Contractor* has included within his quotations for the proposed changes to the Prices and delay to the Completion Date (as planned Completion is delayed). In accordance with clause 62.2 the *Contractor* has included alterations to the Accepted Programme in his quotation.

Should the *Contractor* base the proposed change to the Completion Date on (a) the Accepted Programme at the time of issue of the *Project Manager*'s instruction and request to submit quotations, or (b) the Accepted Programme at the date of the *Contractor*'s alternative quotation submission?

This is one of the problems with delaying the process; things have now moved on. The *Contractor* should have used the Accepted Programme as (a), because that was what was known at that time (clause 62.2).

103. Accepting quotations in a timely manner (clause 62.3)

We are the *Contractor* under an ECC Option C contract. We have submitted a number of quotations to the *Project Manager* and we have agreed to relax the response period for the *Project Manager* to accept our quotations. We now find ourselves in a position in which we are agreeing quotations with our Subcontractors before we have received acceptance of our quotations. We have a number of occasions in which the *Project Manager* is assessing compensation events at a lesser value than we have already agreed with the Subcontractor. What should we do? Also, if a Subcontractor's actual Defined Cost is greater than how the *Project Manager* assesses our compensation event, is the difference a Disallowed Cost?

The first point is to question why you made such an agreement with your *Project Manager*. If it causes such problems, then insist the *Project Manager* deals with your quotations within the time scales set out in clause 62.3. If the *Project Manager* fails to do so then notify him or her of that failure under clause 62.6.

If the *Project Manager* continues to fail to reply for 2 weeks after that notification, then your original quotation will be treated as accepted (clause 62.6) and therefore implemented (clause 65.1), after which only the *Adjudicator* can change it. The timescales in the ECC are deliberately written so that you should not need to get yourself in this position.

The answer to the second question is that this excess amount is not a Disallowed Cost, simply because it is not stated within the list of Disallowed Cost items stated in clause 11.2(25).

104. Failing to respond (clause 62.3)

Our *Project Manager* on an Option A ECC bypass project constantly fails to respond to compensation events that we have quoted. What should they do on receipt of a quoted compensation event and what are the remedies on the contract if they do not?

In order to avoid the indecision and drawn out 'final accounts' of old, the ECC contract is very explicit in this respect.

On receipt of the *Contractor*'s quotation the *Project Manager* has four options only (clause 62.3). Within 2 weeks they are required to accepted the quote, request a revised quote (stating reasons as per clause 62.4), notify that a proposed instruction will not be given or decide to make their own assessment. It needs to be one of those four decisions within 2 weeks.

Failure of the *Project Manager* to respond to the *Contractor* may (discretionary) result in the *Contractor* notifying them of their failure (clause 62.6). Failure of the *Project Manager* to respond within 2 weeks of this is then treated as acceptance of the quotation. This is not revised (clause 65.2). The ultimate remedy in the contract is that continued failure of the *Project Manager* to respond will result in the compensation event being 'implemented'.

The *Project Manager* and *Contractor* should really be working closely in order to gain agreement before submission and resolve queries. Regular meetings are a practical way of keeping on top of the process.

Compensation events

105. What programme should we use when assessing the impact of compensation events? (clause 63.1)

In assessing the impact of a compensation event on planned Completion, the ECC states the latest programme is to be used. How do we work out what is the date of the latest programme? Is it the date when the *Contractor* became aware of the event itself, or the date when the *Contractor* notified the compensation event?

The answer to your question lies in the last part of clause 63.1. If it was a compensation event that the *Project Manager* should have notified (see clause 61.1 for these events) then it is the Accepted Programme that is relevant when the *Project Manager* should have notified, which is when he or she issued the relevant instruction. For all other compensation events, it is the Accepted Programme that is relevant to when the *Contractor* notified the compensation event. You need to bear in mind with the latter that they are subject to an 8 week time bar in clause 61.1.

106. Using the SSCC for Subcontractor's costs (clause 63.1)

Our queries relate to a project using the ECC Option A. In a response to an earlier query on compensation events involving Subcontractors you stated that when valuing compensation events you calculate what the components in the SSCC have cost the Subcontractor, not what the *Contractor* may have paid his Subcontractor. Therefore, assuming that the *Contractor* has used the NEC subcontracts, you will pay what you think the Defined Cost of the Subcontractor should be (using the various rates and percentages in the main contract not the subcontract), excluding any Fee that the *Contractor* has paid his Subcontractor. So, in Option A (incidentally Option B as well) the *subcontracted fee percentage* should be higher than the *direct fee percentage*, because not only does it have to cover the *Contractor*'s overheads and profit, it also has to cover the Subcontractor's overheads and profit.

We have explained this to the *Contractor* and he has stated that having read the contract he can find no such wording to back up this statement, and as such he will be submitting quotations based on the SSCC using the Subcontractor's percentages contained within their Contract Data part two documents and then adding the *subcontracted fee percentage* included within his Contract Data part two document to the totals. In addition, the *Contractor* has also stated that he cannot see how this can possibly be the case as at tender stage he will have many different levels of fees submitted by his Subcontractors, and in some cases he may not even have Subcontractors on board for certain work packages. It would therefore be impossible for him to insert just one Fee. We have responded that he would be expected to make a commercial decision as he would normally do with any tender he submitted.

Our questions are:

1 Do you have any thoughts on this?
2 What process should be followed if the *Contractor* has not used the NEC subcontracts?

For question 1, the *Contractor* needs to read and understand clause 11.2(22), specifically the words 'whether the work is subcontracted or not'. You therefore assess all compensation events using the SSCC, whoever actually undertakes the *works*. You are trying to determine cost to the Subcontractor, not what the *Contractor* has paid or will pay the Subcontractor.

The only reason for having two Fee percentages is to allow the *Contractor* to include for their Subcontractors' fee percentages on work that they subcontract. In Options A and B therefore, the *subcontracted fee percentage* the *Contractor* quotes should be higher than the *direct fee percentage*, because it should allow for both the *Contractor*'s and the Subcontractor's overheads and profit. And, as you rightly say, that requires the *Contractor* to make a commercial decision at tender stage as to the level of the Fee percentages they quote. The *Contractor* could potentially list a number of *subcontracted fee percentages* in Contract Data part two for various Subcontractors, if in turn their mark-ups are so diverse.

With regard to question 2 you follow exactly the same process, in that you have to forecast what the affect has been on Defined Cost. If the *Contractor* has chosen not to use an NEC subcontract that means they may be paid on an entirely different basis to that which they pay their Subcontractor, which may or may not be beneficial to them. That is their concern not yours. You (and the *Contractor*) must forecast what it cost by using the rules in the SSCC.

107. Forecasting the Defined Cost (clause 63.1)

How should compensation events be valued? Is it based on the actual cost incurred by the *Contractor* plus Fee? What happens if the estimate is wrong – is it revised based on the actual?

This aspect of the contract causes a lot of confusion in practice.

Clause 63.1 states that the compensation assessment is based on a mixture of actual Defined Cost of the work already done, forecast of the work not yet done plus the resulting fee.

Note that Defined Cost is a defined term and therefore means different things in different main Options. The 'switch' between actual and forecast is not when the individual assessing gets around to doing it. It is based on the last paragraph in clause 63.1.

If the compensation event results from an instruction, certificate, change in earlier decision or correction of an assumption from the *Project Manager* or *Supervisor* then that communication divides actual versus forecast (note the compensation event may be notified after this point). The last sentence states that in all other cases the date of the notification of a compensation event becomes the switch from actual to forecast – so as soon as the *Contractor* notifies a compensation event, this would require the assessment all to be a forecast.

Compensation events under the ECC are therefore based on largely a forecast/estimate rather than based on actual cost. There are other clauses that need to be taken into consideration, e.g. clause 63.6, which allows the *Contractor* to include time and cost risk allowance. It is all about having a reasoned/fair estimate.

108. Using the *direct fee percentage* (clause 63.1)

We are working with the *Contractor* to agree costs for compensation events and early warnings under the ECC Option A. Under clause 60.1(4), the *Contractor* has notified a compensation event after acceptance of a revised programme by the *Project Manager*. The *Contractor*'s quotation for the compensation event includes Defined Cost for the extended programme, which includes Site accommodation and running costs, key personnel (Contracts Manager, quantity surveyor, Site Engineer, etc.), Site hoarding and so on. After calculating this, the quotation includes a *direct fee percentage* uplift. Can this uplift be used and would not this cost normally be included in the Fee anyway, as per clause 52.1?

We are not sure why you consider that all of these costs should have been included in the Fee because clause 52.1 does not support that view. Defined Cost is a defined term – for Option A see clause 11.2(22). That in turn refers you to the SSCC, which sets out what will be paid for as part of Defined Cost. A compensation event is assessed by its Defined Cost and then the Fee is added (clause 63.1). If it can therefore be shown that one of the costs in the SSCC has been increased because of the compensation event, then it will be assessed in the way set out in the SSCC. Conversely, only if a cost is not listed in the SSCC will it not be part of Defined Cost, and therefore be considered to be part of the Fee under clause 52.1.

So you need to look at the SSCC in detail to decide which of these costs are paid. For example, the cost of people is included in SSCC item 1, as long as their normal place of work is the Working Areas or they are working in the Working Areas. Site-based staff will be paid for and non-Site-based staff will only be paid for when they are on Site. If, for example, the quantity surveyor spends 2 days a week on Site and the other three working in head office then 40% of his Defined Cost is allowed. Equally, from SSCC item 2 the cost of the *Contractor*'s Equipment, including accommodation, is included. You do though need to make sure that it is not already included as part of the *Contractor*'s percentage for people overheads, as set out in SSCC item 41. Therefore, as an example, the Site hoarding will not be paid for because it is part of that percentage, being '... equipment ... for ... security' in the final bullet. The *Contractor* is correct to provide their quotation based on a forecast of what these costs will be – see clause 63.1. In NEC3 contracts you do not wait until the costs have been spent – you forecast what they should be assuming the *Contractor* acts promptly and competently (clause 63.7) and allowing for risks that have a significant chance of occurring (clause 63.7).

> What does worry us here though is that you say that the *Contractor* has notified a compensation event because the *Project Manager* has accepted a revised programme. That cannot be the case. The list of compensation events is set out in clause 60.1, and there is not one that says 'the *Project Manager* has accepted a programme that shows late Completion'. That is putting the cart before the horse! On the other hand, if a listed compensation event means planned Completion is now later than planned Completion shown on the Accepted Programme then the Completion Date will be extended – see clause 63.3. That will, however, be because of the underlying compensation event, not because the *Project Manager* has accepted the subsequent programme.

109. Recovering cost of additional Site staff (clause 63.1)

As *Contractor* and in a compensation event arising in an ECC Option B contract, are we able to claim for additional site staff costs (i.e. Contracts Manager/quantity surveyor) to cover the pricing and delivery of a compensation event, even when the overall duration has not increased?

With regard to the pricing of the compensation event this is excluded – see the last part of clause 11.2(22). You therefore have to make allowance for this cost in your Fee. Compensation events are assessed based on their effect on Defined Cost – see clause 63.1, but remember that Defined Cost is a defined term – see clause 11.2(22) and the SSCC.

In the SSCC you will see you only get paid for people when they are in the Working Areas, and only if your cost of them has increased. Just because your Site Agent or perhaps foreman worked on them does not mean that your cost increased if they were going to be there anyway. They just maybe worked a bit harder. On the other hand, if the time you are on Site was extended because of the compensation event then their Defined Cost would increase. Or if the compensation event was such as to require an additional foreman to run it, again the cost of that would be included. Just because somebody works on it does not mean your cost increases.

110. Charging for staff time in a compensation event (clause 63.1)

On an ECC Option A contract, I have the following three questions to ask:

1. Can a *Contractor* charge for his quantity surveyor's time in preparing a compensation event quotation?
2. Can a *Contractor* charge for his quantity surveyor's time in sorting out quotes and placing orders necessary for a compensation event quotation?
3. Can a *Contractor* charge for his Site foreman's time in working on a compensation event in his quotation for a compensation event?

The answer to your first question is no because clause 11.2(22) specifically excludes the cost of preparing quotations for compensation events.

The answer to your other two questions is a bit more complicated, and it will depend on the circumstances. The value of the compensation event is based on its forecast effect on Defined Cost – see clause 63.1. So the *Contractor* has to show that the Defined Cost for these people increased. If they worked on them and were going to work on Site anyway, then the *Contractor* cannot show that the Defined Cost for them has increased. All that has happened perhaps is that the people have worked a bit harder. On the other hand, if the compensation event delays Completion by, say a week, then the *Contractor* and these people will be on Site for another week and their Defined Cost will increase, and that increase is included in the valuation of the compensation event. Equally, if the compensation event is of such a size that it is forecast that an additional foreman will need to be brought in to manage it then again that will increase Defined Cost, and that will be included in the valuation of the compensation event. So the answer to the second and third question is 'it depends'.

111. People costs in a compensation event quotation (clause 63.1)

As part of an ECC Option A contract, when assessing compensation events, the *Contractor* is including time for their quantity surveyor who is based in their head office (so outside of the Working Areas) and their senior engineer who is based full time at the Site. The *Contractor* is tending to charge a nominal amount of 2 hours for the quantity surveyor and 4 hours for the senior engineer for every compensation event. Please confirm if the *Project Manager*'s following interpretations of the contract are correct.

1. Under SSCC item 1 people, the first category of people excludes *Contractor*'s staff, who are working at the head office. Therefore the *Contractor* cannot claim time for the quantity surveyor as his time is covered by the Fee.
2. Under SSCC item 1 people it includes people whose normal place of working is within the Working Areas, so the senior engineer is a role that can be charged. However, in this instance the senior engineer is already assigned full time to the contract so the *Employer* is already paying for 100% of his time. So, if the senior engineer is directly working on an activity that is the subject of a compensation event, the *Contractor* can claim for his time.
3. If the senior engineer is not directly working on an activity that is the subject of a compensation event but the compensation event delays planned Completion, the *Contractor* can claim the cost of his time for the delay period only.
4. If the senior engineer is not directly working on an activity that is the subject of a compensation event and the compensation event does not delay planned Completion, the *Contractor* cannot claim the cost of his time.

The assessment of a compensation event is based on what effect it is forecast to have on Defined Cost. The answer is not quite as simple as you have set out. The first thing to say is that the cost has actually to increase. If the *Contractor* has an engineer or quantity surveyor on Site anyway and he uses him to manage the compensation event then his Defined Cost has not increased, so he is not included. Of course, if the compensation event is such as to lead him to need another engineer on Site (either because it is large in itself, or there have been a lot of other changes and this one 'breaks the camel's back'), then that will be included. Or if it delays his work so that

> the engineer is on Site for longer than he would have been, then that will be allowed for. With regard to people working outside of the Working Areas, they are not included as part of the definition of Defined Cost, and, under clause 52.1, they are assumed to be included in the Fee. So your interpretation 1 is correct; 2 is only correct to the extent that his cost can be shown to be forecast to increase; 3 is correct, as is interpretation 4.

112. Are compensation event quotations based on actual cost? (clause 63.1)

Our queries relate to a project using the ECC Option A and are:

1 Clause 63.1 states 'The date when the *Project Manager* instructed or should have instructed the *Contractor* to submit quotations divides the work already done from the work not yet done.' If the *Contractor* does not submit a quotation until well after the *works* are complete should the quotation be based on the actual cost or on forecasts of the actual cost of the *works*?

2 If the *Project Manager* does not instruct the *Contractor* to submit quotations until after additional *works* were carried out, due to them not being brought to his attention by the *Contractor* until after they were complete, should the quotation be based on the actual cost or on forecasts of the actual cost of the *works*?

The answer to question 1 is that you assess using a forecast, no matter when that assessment is being done. And that forecast assumes the *Contractor* acts promptly and competently (clause 63.7) and allows for risks that have a significant chance of occurring (clause 63.6).

With regard to question 2 the answer will depend on the reason for the compensation event. If it arises from a *Project Manager*'s instruction to change the Works Information, then the obligation is on you to notify the compensation event and instruct a quotation (clause 61.1). So you will use a forecast because it is from the date you 'should have' instructed a quotation. If it is one that the *Contractor* should notify, then you first have to consider whether the time bar in clause 61.3 applies. If it does not then you use actual Defined Cost, but again by assuming the *Contractor* acted promptly and competently.

113. Compensation event quotations and Subcontractors (clause 63.1)

We are using ECC Option C and have a quotation for a large compensation event for additional steelwork, which we are worried is excessive. We are looking for confirmation on using a transparent method to review the proposed costs and accept them (or not) in accordance with the contract. The quotation includes both costs for manufacture and large amounts for Site labour, which we cannot see any justification for. We understand the vast majority (if not all costs) stem from the steelwork fabrication Subcontractor, who will both supply the steelwork and fit it on Site. The quotation includes large sums under people costs. Looking at the SCC descriptions and guidance, we were thinking this should only be *Contractor*'s directly employed staff, not subcontracted people. We do note the SSCC people costs do, however, allow for Subcontractor people costs on Site.

Should the *Contractor* be using SCC for quotations involving Subcontractor's costs or should we be agreeing to use SSCC to break down all compensation events involving Subcontractors? Or should they simply give me the Subcontractors' quotations to accompany their own quotation and add on their Fee percentage for Subcontractors? If it is the latter, how can we have sight of where the costs are going so we can consider whether the costs are justified? Can we ask that the steelwork fabricators quotation be broken down – and if so to what level of detail?

This compensation event is valued using Defined Cost plus Fee – see clause 63.1. The definition of Defined Cost in ECC Option C is set out in clause 11.2(23). Here you will see that for subcontracted work it is based on what the *Contractor* is forecast to pay his Subcontractor, to which is added the *subcontracted fee percentage*. So you simply do not use the SCC at all to value the subcontracted element of Defined Cost of this compensation event. The definition of Subcontractor is set out in clause 11.2(17). The only work that is valued using the SCC is work that is not subcontracted. In addition, you only use the SSCC if you both agree to do so or if you want to when you make your own assessment (clause 63.15); otherwise you use the SCC not the SSCC. From the question we are having difficulty understanding exactly what work is subcontracted and what is not, so it is difficult to comment in detail. If, for example, the fabrication and erection work was being carried out by one Subcontractor, then you do not use the SCC at all for that element. On the other hand, if the steel was being fabricated by one supplier and erected by a (different) Subcontractor then you would use section 3 of the SCC for the fabrication and then the amount that the *Contractor* pays his Subcontractor for the erection. As to the amount the *Contractor* pays its Subcontractor, that will depend on the terms of the subcontract they are under, which you should have seen when accepting that Subcontractor – see clause 26. If, for example, the subcontract values variations based on rates then those rates are used and both the *Contractor* and you are stuck with them and cannot ask for any further substantiation. On the other hand, if the contract is based on assessing Defined Cost then you are both stuck with that.

In all of this it is necessary for the *Contractor* to provide enough information to satisfy you that that is what they are going to pay. Remember this is an 'open book' contract so you are entitled to see whatever the *Contractor* has on that (clauses 52.2 and 52.3). You are also entitled to see, in the case of a subcontract value based on Defined Cost, that that assessment has been carried out at open market or competitively tendered rates (clause 52.1). That may well require a detailed breakdown. It is not possible to tell you how much information you will need in order to do that, as it will depend on the circumstances. This is something that you need to discuss with your *Contractor*.

114. People not working in the Working Areas (clause 63.1)

Q

The Working Areas in our ECC contract is 'the Site'. We are applying for compensation events as per the SSCC, which includes project managers, quantity surveyors, buyers and so on. The *Project Manager* is stating that as these individuals are working from head office and not 'the Site' they are included within the Fee and cannot be separately charged for.

Our queries are, first, as these individuals are listed in the SSCC, can they be charged for even if they are not working in the Working Areas? Second, it is my understanding that the process with compensation events is to put you in a position that you would have been in had the compensation event never occurred in the first place. Therefore, I believe I should be entitled to charge for individuals who assisted in preparing quotations for and undertaking compensation events.

A

The *Project Manager* is correct for two reasons. The first is the one he has given you, people are only paid for when and if they are working in the Working Areas – see the bullets at item 1 of the SSCC. Any other costs are assumed to be in your Fee – see clause 52.1.

In answer to your second question, it is intended to be neutral, but it assumes that you have allowed enough in the Fee you quote to cover what you are not going to get paid for. If you have under-priced the Fee then this will be at your risk. The second reason is that the value of compensation events will be based on what your Defined Cost has increased by – see clause 63.1. Just because an employee works on something does not necessarily mean that has increased your Defined Cost if you would have employed him or her anyway.

115. Extensions of time? (clause 63.3)

How are extensions of time claims dealt with in the ECC contract? Do you compile these at the end of the project when you know the overall impact of all the variations?

In short – absolutely not! The ECC is very different to traditional forms of contract. Each compensation event needs to be assessed within professional timescales and both the time and cost impact are assessed jointly. The *Contractor* has 3 weeks to assess both (clause 62.3) or a later period agreed with the *Project Manager* (62.5).

The latest Accepted Programme is adopted (clause 63.3), then updated for remaining work (clause 62.2). We then assess whether planned Completion (i.e. when the *Contractor* plans to complete) is affected. If it is affected, then by whatever duration we also add this to the Completion Date. The ECC provides very clear rules for assessing the programme impact within the core clauses. There is no room for assessing at the end of the project – clause 65.2 clearly states that assessments are not revised if a forecast is wrong – this is applied to both time and cost so we have a 3-week window to assess and ensure this is accurate.

116. Dominant cause of delay (clause 63)

A design and build project is being carried out under ECC Option A. The project is linear involving access onto adjoining green field lands owned by different people to construct a roadway; although there are *sectional* Completion Dates there is only one *access date* included in the contract. The clause 60.1(13) compensation event for weather has been deleted.

The project has suffered an overall delay of 10 weeks. Firstly, 3 weeks of this delay was experienced due to inclement weather. Shortly afterwards, the *Contractor* progressed the *works* such that access was required from one piece of land onto another. The *Employer* advised the *Contractor* to demobilise, as agreements had not been fully concluded with a particular land owner, and centred around more land-take now required than had previously been planned. Due to this, a further delay of 7 weeks followed where the *Contractor* was in delay due to the stand-down.

The *Contractor* maintains that had there been no weather delay the non-access delay would have been realised earlier, yet still not resolved until the time it actually was. The basis for this point of view is the fact that the *Contractor*'s design drawings, and revisions thereto at the request of the *Project Manager*, were submitted in good time (and accepted) well ahead of the planned start of *works* onto the affected lands; therefore the *Employer* had every opportunity of agreeing any additional land-take required with the land owner in good time to obviate the subsequent delay and ahead of the weather delays. The *Contractor* maintains that the dominant delay is the non-access. Currently, the *Project Manager* recognises the 7-week delay only as a direct effect of the non-access, but is holding the *Contractor* liable for the 3-week weather delay and delay damages.

Is the *Project Manager* correct in his approach and what is the protocol for establishing the dominant delay?

Delays that the *Contractor* knows about have to be dealt with in each programme that the *Contractor* issues for acceptance. We assume here that they are to be submitted monthly. The *Contractor* therefore should have included in its next programme issued for acceptance the effect this 3 weeks has had on the actual progress and the order and timing of the remaining work. That may show planned

Completion being moved back by 3 weeks or it may show the remaining *works* being re-programmed to catch up the 3 weeks. Once this programme is accepted it becomes the Accepted Programme (clause 11.2(1)) and is used to assess future compensation events, including this one.

If you did this, then the answer will depend as to what happens when this further delay is added to the programme and the answer will normally be, given the facts you have set out, that the date for planned Completion will be moved by 7 weeks. In that case the Completion Date will be moved by 7 weeks (clause 63.3). That will be the same whichever way you dealt with the initial 3-week delay in your programme. And if you did not provide the programme for acceptance that you should have then the *Project Manager* can reject your assessment of the compensation event (third bullet of clause 64.1) and make his own assessment based on what he thinks the programme should be (clause 64.2), including, of course, the effect of the first 3-week delay.

Delays are therefore dealt with in the order that they are known about. The term 'dominant delay' comes from the case law on other contracts, which have very different requirements as to extensions of time. Even with those contracts the law would say that both are 'dominant' at different times, so the outcome would be the same. The *Project Manager* appears correct to assess this delay due to the compensation event as being 7 weeks, not 10 weeks.

117. How does the contract deal with multiple programmes that have not been accepted? (clause 63)

During the execution of an ECC Option C contract, programmes were submitted on a regular basis and these were accepted by the *Project Manager*. However, the *Employer* changed the *Project Manager* and over the next 5 months no programmes were accepted. Most of the revised programmes were ignored and others not accepted for reasons not stated in the contract under clause 31.3. The contract does not appear to deal with this scenario other than, in clause 60.1(6), to state that no reply by the *Project Manager* within the period required is a compensation event.

In order to agree the effect that each compensation event had on planned Completion, a delay analysis has been produced, by working from the last Accepted Programme, inputting the time impact of the 'critical path' event, to establish a revised programme and planned Completion. This new programme (although not accepted) is then used to assess the delay of the next compensation event and so on, forming a step-by-step process and a series of non-accepted programmes with the delays clearly identified and the new planned Completion demonstrated. These delays are identified within the appropriate compensation event quotation. Although this situation should not have arisen, can you advise how the contract envisages such a situation will be dealt with?

If the *Project Manager* fails to reply within 2 weeks to the programme, that will be a compensation event under clause 60.1(6). Equally, if the *Project Manager* does not accept the programme for a reason not stated in the contract, that will again be a compensation event, this time under clause 60.1(9).

So the contract deals with both scenarios and you should have notified these compensation events at the time and in accordance with the contract. If there is a dispute between you as to what the Accepted Programme was or should have been then the *Adjudicator* will have to decide that. The *Adjudicator* may well start from the premise that the programmes you submitted at the time were correct, as they were not rejected for reasons within the contract by the *Project Manager*. It is important to understand that clauses 63.1 and 63.3 require that the assessment is made using forecasts made with the foresight the Parties had at the time, not the hindsight they have now. So you use the Accepted Programme to assess the compensation event, then revise the Accepted Programme (at the intervals in the contract) to take into account all of the delays up until then, including your own, as well as the compensation events) to come up with what the Accepted Programme should then have looked like, and so on. So it is not just a case of feeding the delays caused by compensation events in each month, you need to allow for your own delays, etc. That is our view as to how the matter should be dealt with. However, it does not follow that every adjudicator would agree, and for that reason we are unable to offer detailed advice and suggest that you should get further legal advice if you are still unsure.

118. Including risk in compensation events (clause 63.6)

Q I am a *Project Manager* on an ECC Option B project. My question is about compensation events and whether risk should be included in the *Contractor*'s quotation? If so, where does it say this in the contract?

A Under clause 63.6 we need to assess both time and cost allowances for risk within each compensation event. The clause states that they need to have a significant chance of occurring and of course be the *Contractor*'s risk in the contract. We tend to find that if the *Employer* has amended the compensation event section by way of Z clauses then they may get a nasty surprise here as the compensation events, quite rightly, will be more expensive and include more time risk allowance (as the *Contractor*'s risks are greater).

Pouring a concrete slab in the UK in December in an exposed location would naturally include time and cost risk allowance to account for the impact of bad weather. If the concrete pour was internal/protected then this would not be appropriate. This aspect is best discussed between the *Project Manager* and *Contractor* in order that they can understand one another's logic and thought processes and ensure that compensation events are managed in a timely fashion. The way we look at this is that we have a duty to our respective organisations to manage professionally and inform them of the time and cost impact of compensation events.

119. Inconsistencies in the *Contractor's* Works Information (clause 63.8)

On our hospital project under ECC Option C all of the design responsibility is with the *Contractor*. We have signed the contract but since discovered that their building drawings contradict the mechanical and electrical details in a number of instances. For example, one drawing shows three light pendants and another shows five. We want five and the *Contractor* is telling us that would be a compensation event – where do we stand with this?

The answer is quite simple and is contained in clause 63.8. Presumably, in the Works Information you have clearly stated which aspect of the design the Contractor is responsible for (as per clause 21.1). We therefore have an inconsistency – this should be notified under clause 17.1. We then instruct the *Contractor* to correct their Works Information (clause 14.3) in order to correct the inconsistency and the answer to who it is found most favourable to is contained in clause 63.8.

It is found most favourable to the Party that did not provide the Works Information, i.e. you. So, in contrast to being a compensation event, they proposed three and elsewhere five (their inconsistency); if you want five then there is no time and cost impact. If you actually wanted three then it would be a negative compensation event.

This works the other way because any inconsistency or ambiguity in the Employer's Works Information is found most favourable to the *Contractor*.

The moral of the story is therefore (before contract signature) to ensure that one another's Works Information is free of ambiguity and inconsistency.

120. Unclear Works Information (clause 63.8)

An ECC Option A contract was tendered in which there were a number of reinforced concrete headwalls. There were no reinforced steel bar bending schedules accompanying the tender documents – a small reference box was included on the reinforcement tender drawings stating 'estimated reinforcement quantities for structures shown on this drawing for tender purposes only – 164.71 tonne'. The contract was awarded in September with a starting date of November. The contract documents state that the steel bar bending schedules are included in the contract, but will be provided on award of contract; the contract drawings still contain the reference box with the estimated quantities.

The *Project Manager* changed the Works Information and notified a compensation event in February stating that 'Estimated steel quantities were used at tender stage of the contract, the actual amount of reinforcing required is different from the tendered amount.' The net effect is that the amount of reinforcing has been reduced by 38.21 tonnes.

Would estimated quantities on a tender drawing be deemed to be Works Information?

The tender drawings should be exactly the same as the contract drawings. If you are later asked to enter into a contract based on Works Information (including of course the drawings) that is different from that which your *activity schedule* is based on you should refuse, as that could be a major problem! If you are not careful, the formal contract will be whatever you agree to at the time, and you might be stuck with a different Works Information and *activity schedule*. In this case, for reasons below, this does not appear to be the case here. The answer to your question is 'yes', by the way, as it is the contract Works Information that you will have assumed to have taken into account when pricing the work. So, what exactly was in the Works Information in the contract – did it include both the drawings and the bending schedules or were the bending schedules issued afterwards? If the answer is that both were in the contract Works Information then that is an inconsistency that the *Project Manager* should issue an instruction to resolve (clause 17.1). Presumably the *Project Manager* instructed you to use the bending schedules? If so, that was a change to the Works Information, in the sense that the *Project Manager* instructed you to ignore the drawings. That is a compensation event under clause 60.1(1). And that compensation event will be valued using the principle set out in clause 63.8. As the Works Information was provided by the *Employer* then it is assessed in your favour, i.e. you are assumed to have allowed for in your price the lower quantities set out in the bending schedules. There is therefore is no reduction and the compensation event has no effect on cost or time.

On the other hand, if the bending schedules were not in the contract Works Information then your price is assumed to be based on what is on the drawings – after all there is no inconsistency is there? When the *Project Manager* subsequently issues the revised schedules, that will be a change to the Works Information, which is again a compensation event under clause 60.1(1). And it will be valued using the principles in clause 63.1, i.e. in general terms, how much did the change affect the Defined Cost for carrying out the work plus the Fee? And if that is a negative figure (as it is in this case) the total of the Prices is reduced (see clauses 63.2 and 63.10).

121. *Contractor* not assessing compensation events (clause 64.1)

Q

We are becoming really frustrated on our ECC Option C project as the *Contractor* is just not assessing compensation events within the 3-week period. In some instances we have extended the timescales under clause 62.5 in order to give them more time but they still fail to assess.

The *Employer* is increasingly annoyed as they want to know where they stand. What provisions are in the contract for dealing with this?

A

OK, the contract is clear on how this is dealt with. Clause 64.1 provides four reasons/scenarios in which the *Project Manager* must step in and make their own assessment of a compensation event.

Note the opening sentence states the *Project Manager* assesses, it does not say 'may' (which is discretionary). So you have to step in and make your own assessment – this is your obligation in the contract. It is very important to note that you need to act as stated in the contract and work in a spirit of mutual trust and co-operation (clause 10.1). Even if it is now up to you to assess we would still be seeking to meet with the *Contractor* and ensure the assessment is accurate. You also need to take account of other assessment clauses in the contract.

122. When is a compensation event 'implemented'? (clause 65.1)

When does a compensation event become agreed/accepted under an ECC project?

We have a clearly defined process for managing compensation events under the ECC (i.e. change control). This consists of a four-stage process: notify, quote, assess, implement. It is important to use the correct terminology – we do not agree/accept as such. A compensation event achieves the status of having been implemented based on three potent reasons as defined in clause 52.1.

The three reasons for a compensation events being implemented under clause 65.1 are: the *Project Manager* accepts the *Contractor*'s quote, the Project Manager notifies his own assessment, or the *Contractor*'s quotation is treated as having been accepted (refer to clause 62.6).

If any of those three scenarios occurs then it has achieved the status of being 'implemented'.

123. What is an implemented compensation event? (clause 65.2)

Can you please define what an implemented compensation event is under ECC, clause 32.1? For example, does it mean when it is acknowledged by the *Project Manager* as the *Employer*'s liability, or when a quotation has been agreed? Would it be reasonable to suggest that liability for a compensation event could/should be agreed at on-going risk reduction meetings allowing the programme to be updated regularly and remain meaningful, even though the compensation event may not have been priced? If you agree with this approach how ideally should this agreement be recorded?

The way that compensation events are implemented is set out in clause 65.1 and the consequences in clause 65.2. Early warnings and risk reduction meetings have nothing directly to do with compensation events. They are about matters that may occur (i.e. a risk) in the future, and are about managing that risk, whoever's it is. Compensation events are about things that have happened or will happen and are at the *Employer*'s risk, because they are listed as a compensation event. Not all early warning notifications are about future compensation events and not all compensation events can have an early warning.

We recommend that you keep the processes for dealing with risks of things that may happen separate from dealing with compensation events. And again updating programmes is something different. The programme has to show both the Completion Date (contractual) and planned Completion (practical) – see clause 31.2. The latter is linked through the activities in the programme, the former is not. The latter can move for all sorts of reasons, the former can only move because of a compensation event. If you know something is going to happen that you think is a compensation event then its effects go on the programme straight away and affect planned Completion. However, it is not until the compensation event is implemented that the Completion Date shown on the programme can be moved.

124. Having your cake and eating it! (clause 65.2)

We have a *Project Manager* on our ECC Option C contract who, every time our cost is lower than the compensation event quotation, is assessing based on actual cost. Conversely, when our costs exceed our quotation he uses the quotation. The term 'having your cake and eating it' springs to mind here.

What is the correct manner of assessing compensation events and which specific clauses do we need to refer to.

There are a number of clauses that define how compensation events should be assessed. Clause 63.1 states that this is largely a forecast of cost; clause 63.6 ensures that time and cost risk allowances are included (as appropriate); clause 63.7 also states that a compensation event is based on the assumption that the *Contractor* reacts competently and promptly for example.

The key clause in this question is clause 65.2. This states that a forecast (both time and cost) is not revised if a forecast is found by later recorded information to have been wrong.

The compensation events process under the ECC ensures that a reasoned estimate becomes the basis of the compensation – not what the actual cost/time were. The actuals are academic and not relevant – it is all about having a reasoned estimate and that becoming the basis of change to the Prices, Completion Date and Key Dates as appropriate.

125. Can you change an accepted quotation for a compensation event? (clause 65.2)

Q Given that the intent in ECC is that a compensation event is quoted for, the quotation is agreed and the work carried out and the issue therefore closed, is there any provision for amending a quotation if it is found later to contain significant costs that should not have been accepted?

A The first thing to say is that most compensation events should be assessed on a forecast of Defined Cost, not actual Defined Cost – see the last part of clause 63.1. With that in mind, the answer to your question is no, it cannot be revised. Once the *Project Manager* has accepted a quotation for a compensation event it is implemented – see clause 65.1. It cannot subsequently be revised if the forecast on which it is based is later shown to be wrong – see clause 65.2.

Some further thoughts and top tips

- Compensation events do not manage themselves – agree at project start how these will be managed. Consider a fortnightly meeting to keep on top of this.
- Follow the process in the contract. Do not skip stages.
- Include the planner – both cost and time need to be assessed jointly. Time is money!
- Remember clause 63.1 – compensation events are largely a forecast. Time and cost risk may be included.
- Use the ability to make *Project Manager* assumptions under clause 61.6. This is an excellent means of allocating risk and obtaining better value for money.

Chapter 7
Title

126. Does title mean ownership? (clause 70)

Q We have been advised that the ECC does not contain entitlement to payment for Plant and Material materials on or off Site. We note clause 70.2 states that title passes to the *Employer* for Plant and Materials brought within the Working Areas. Does title mean ownership and if so, how can this pass if no payment is made?

A Things are never quite as simple as they seem and title is not quite the same as ownership. Payment and title do not have to be directly linked as title can pass without payment being made and payment can be made without title passing. It will all depend on the terms of the contract the Parties have entered into. In the case of the ECC title can pass without payment being made. It is, by the way, the same with most standard form construction contracts as they will generally state that title will pass as soon as such material arrives on Site. You will often find though that payment is actually rarely made in advance or at that time and will often be made some time afterwards. Even then it will often only be at 80% or so until the materials are actually incorporated into the *works*.

For ECC, it is also not the case to say that in all Options the *Contractor* will not be paid for Plant and Materials on Site. In Option A, it will depend on how the Activity Schedule has been completed. For example, if there is an activity that says 'Item x delivered to Site' then the *Contractor* will be paid that amount in the assessment after item x arrives. Equally, if an item in the Activity Schedule says 'Complete manufacture of item x' then when it is complete the *Contractor* will be paid in the next assessment even if the Plant and Materials are not on Site. See clauses 70 and 71 for the safeguards in that case. With Options C, D and E the *Contractor* will often be paid Defined Cost before it actually pays for such Plant and Materials, and that will include for Plant and Materials wherever they are – see clause 11.2(29 and (23)) and item 3 of the SCC.

127. Vesting Plant and Materials (clause 70.2)

On an ECC Option A contract, does the inclusion of clause 70.2 avoid the need for vesting documentation for Plant and Materials on Site if the area for storage is included in or added to the Working Areas under clause 15.1?

We are limited in what we can say because the question of vesting certificates is a legal question. Once any Plant and Materials (not Equipment) are brought within the Working Areas whatever title the *Contractor* has to it passes to the *Employer* (see clause 70.2). However, the law around this is always complex and, as a general rule, the *Contractor* cannot pass to you title it does not have, whether by the operation of clause 70.2 or by the signing by the *Contractor* (as opposed to the title holder) of a vesting certificate.

With regard to storage it is important to understand that the definition of the Working Areas can only include areas that are being used only for work on the contract (see second bullet of clause 11.2(18) and also clause 15.1). Therefore, the area will and should be under the direct control of the *Contractor*. The more remote from the *works* the Working Areas are, the more risk it is to the *Contractor* and *Employer* when it comes to Plant and Materials being removed for whatever reason. This is a risk that the *Employer* will need to manage.

128. Title to materials from excavation (clause 73.2)

On an ECC Option D contract, who has title to surplus materials from any excavation works?

Clause 73.2 states that the *Contractor* has title to materials from excavation and demolition only as stated in the Works Information. The default therefore will be that it will be the *Employer* that has title to such materials unless stated otherwise in the Works Information. This will of course demand careful thought at tender stage to decide what is best for the particular project.

 **Some further thoughts and top tips**

- Make sure when drafting the ECC Works Information that it contains an appropriate statement if you wish the *Contractor* to have title to materials from excavation and demolition.

Chapter 8
Risks and insurance

129. Weather ... it's a compensation event (clause 80.1)

Q I am trying to understand risk under ECC3. Clause 80 seems to be quite clear as to who bears risk for what. Assuming that the *Employer* does not amend Contract Data part one to add to the standard *Employer*'s risks under clause 80, why for example does an abnormal weather event become a compensation event under clause 60.1(13)? Surely, if weather is not included in Contract Data part one, then it is the *Contractor*'s risk – full stop? So how can there be compensation?

A It is important to understand that the risks in clause 80 are merely risks to physical damage and third party responsibilities. They do not cover the financial risks for carrying out the work itself. That financial (and time) risk is set firstly by the list of compensation events. Compensation events are always at the *Employer*'s risk. The other major factor with regard to the remaining financial risks (although not time risks) for carrying out the *works* is the choice of main Option. With Options A and B most of the financial risks for carrying out the *works* lie with the *Contractor*. In Options C and D most of them are shared between the Parties. In Options E and F most of them lie with the *Employer*.

130. A hypothetical situation (clause 80.1)

Clause 80.1 states that 'The following are the *Employer*'s risks. Claims, ... and costs payable which are due to: use ... of the Site ... for the purpose of the *works* which is the unavoidable result of the *works*'. Would it be correct to say that in a hypothetical situation, if a contract required a large number of heavy goods vehicle movements to import a very large volume of stone, and if the 3.5 m access road that formed the access to the Site, and was within the limits of the Site, started breaking up, that the costs associated with the remediating of the access road would fall within the coverage of the stated part of clause 80.1 and hence be the *Employer*'s responsibility?

Nice try but ... no! 'Unavoidable' means that it could not be avoided no matter what you did, and we cannot see how this damage could be called 'unavoidable'. You could have avoided it by carrying out the stone placing in a different way – using smaller lorries, or dumpers and stockpiles. Or you could have put a protective mat over the road, or you could have constructed a different temporary road and used that instead. All of those would have avoided this damage, so it cannot be said to be 'unavoidable'. All would cost much more, but that is irrelevant. 'Unavoidable' does not mean the same as 'not economically viable'.

131. Copper theft! (clause 80.1)

We have been employed by our client who has undertaken that the contract we are working under is ECS Option A – which is back to back with the *Employer*. The question is that when working on Site we had to leave materials (copper fittings) in the building and I was under the understanding that the site should be safe and secure. There was a theft and I would like to know who is responsible for this?

The first point we need to make is that NEC contracts cannot be entered into just by reference to them. There is an important document called the Contract Data that needs to be completed. In your case, assuming you are using the ECS, the *Contractor* (your client) needs to have filled out the Contract Data part one and you need to have filled out the Contract Data part two. Without that information the 'contract' is simply too uncertain to be valid. Assuming that you do have a valid ECS then the answer is normally that you are entirely responsible for the security of the Plant and Materials (a defined term) left on site and you have to insure against their loss (see the Insurance Table in clause 84.2). Under the ECS this is therefore your risk – see clauses 80.1 and 81.1. You therefore have to look to your insurers to recover these costs. The only possible exception is that if the theft was as a result of the fault of the *Contractor* (your client) or the *Employer* (the *Contractor*'s client), for example if they had left the Site unlocked at night when they should not have. Then that may be the *Contractor*'s risk – see the third sub-bullet of the first bullet of clause 80.1. If that is the case then it would be a compensation event under clause 60.1(14) of the contract. Compensation events are the only way that you will get paid any more money or allowed any more time in the ECS. The process involved is set out in clauses 60 to 65. It is important for you to understand the processes involved – for example you must notify a compensation event within 7 weeks of becoming aware of the event, otherwise you lose your entitlement – see the last sentence of clause 61.3.

132. Is terrorism a compensation event? (clause 80.1)

Q Within the ECC contract under section 8 – terrorism is not referenced as an *Employer*'s risk. Does this mean that by default acts of terrorism are a *Contractor* risk and something that should be included in the *Contractor*'s insurance policy? Or should the contract be specific and ask the *Contractor* to take out specific insurance cover for acts of terrorism? If no specific request is made then will it be an *Employer* risk and treated as a compensation event under clause 60.1(14) if it occurs?

A As you rightly say the answer to this will be in section 8 of the contract. You are also right to say that terrorism is not specifically mentioned in the list of *Employer*'s risks in clause 80.1. However, there are items such as civil war, rebellion, revolution, insurrection, riots and civil commotion. Whether these cover any or all acts of terrorism is a legal question. We can only suggest that you get some specialist legal advice on this. If this is covered then it is not your risk and you do not need to insure against it. However, if terrorism is not included within clause 80.1 then clause 81.1 of the contract clearly makes it your risk. You will therefore be required to provide against it for the insurances set out in the first two rows of the Insurance Table.

133. Take over versus Completion (clause 80.1)

I am confused between the terms 'take over' and 'Completion' under ECC3. We have a situation in which the *Contractor* achieved Completion one week prior to the Completion Date. Completion was certified but a member of the public damaged the *works*. Who is liable?

The first thing to remember is that take over and Completion are two different concepts and are not directly related. Completion is about the state the *works* are in (clause 11.2(2)), whereas take over is about who becomes responsible for the *works* (clause 80.1). As you rightly say, the *Contractor* remains responsible for the *works* (and has to insure them) until take over, not Completion, has happened. And take over can occur before, at, or after Completion. To add the mix, you also need to remember that the Completion Date is the date by which the *Contractor* is obliged to achieve Completion (clause 30.1). And again, Completion can actually be achieved before, at, or after the Completion Date. Of course, if it is the latter of those three the *Contractor* will be in breach of contract.

The 'normal' position is that take over actually occurs 2 weeks after Completion, whether the *Employer* wants it to or not (see second sentence of clause 35.1). So if the *Contractor* achieves Completion 6 weeks before the Completion Date (i.e. 6 weeks 'early') then take over occurs 2 weeks afterwards, i.e. 4 weeks before the Completion Date. Now some projects may not be convenient for the *Employer* – for example, its staff may not be ready to occupy it early. In that case the *Employer* may, before making the contract, make it clear to the *Contractor* that it does not want to take over the works before the Completion Date. That is done by inserting an optional entry in the Contract Data part one). So the *Contractor* then knows when it is pricing the work that if it finishes 'early' it will still be responsible for the *works* until the Completion Date, and can price that risk accordingly. In that case (and only in that case) the first sentence of clause 35.1 applies.

In addition to all of this the *Employer* may, if it wants to, start to use all or any part of the *works* at any time, which can be before, at, or after Completion has been achieved (first sentence of clause 35.2). If it does so it takes over those parts of the *works* (second sentence of clause 35.2), unless the exceptions listed in the bullets of clause 35.2 apply. So, if the *Employer* starts to use the *works* as soon as Completion occurs, take over occurs at that point. Also, take over before Completion has been achieved will be a compensation event unless it happens after the Completion Date (clause 60.1(15)). Whenever take over occurs the *Project Manager* is required to certify it (clause 35.3). Usually Completion and take over occur on the same date, because the *Employer* is keen to start using the *works*. In that case the *Project Manager* issues two certificates for that day, one for Completion and one for take over.

In your case, assuming there is no optional statement in the Contract Data, if the *Employer* had not started using the *works* and if the damage occurred within 2 weeks of Completion, then the *Contractor* is liable for this damage.

134. How to define take over (clause 80.1)

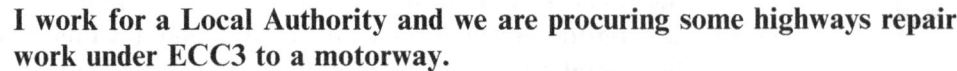

Q I work for a Local Authority and we are procuring some highways repair work under ECC3 to a motorway.

I want to make it clear that the various take overs (as the work is under way by members of the public in order to ensure traffic flow) are not our liability. I want to avoid a situation in which damage caused by a car accident for example could be construed as an *Employer*'s risk. How do I clarify this in the contract documents?

A The answer lies, as it often does, in careful drafting of the Works Information. If you follow the April 2013 guidance on 'How to ... write the ECC Works Information' (which is advisable) then you are directed to the manner in which this should be drafted. WI440 draws attention back to clause 35.2 of the ECC. This states that the *Employer* may use any part of the *works* before Completion has been certified. If so, take over begins when they begin to use it. However, the clause goes on to state 'except' (first bullet) for a reason stated in the Works Information.

So, if you describe the reasons clearly within the Works Information then (as per your aim) the *Contractor* will remain responsible.

135. Insurance of existing buildings (clause 84.2)

Q I am currently drafting an NEC3 ECC form of contract and would be grateful if you could advise whether insurance of existing buildings (to be retained and refurbished as part of the *works*) must be taken out by the *Employer*, or is included within the insurance provisions contained in the Contract Data?

A If you look at the Insurance Table in clause 84.2, the first row of insurance details loss of or damage to the *works*, Plant and Materials. This covers such loss and damage to the new *works* themselves.

The third row of insurance details liability for loss of or damage to property (except the *works*, Plant and Materials and Equipment). This covers, for example, property other than the *works* itself.

The *Contractor* is thus obliged to provide both of these insurances unless otherwise stated in the Contract Data part one (see clause 84.1).

It may be that the *Employer* has an insurance policy in place that would cover the existing buildings and their refurbishment. If the *Employer* has not got such insurance, then the default contractual position is *Contractor*-provided insurance as stated.

136. Damage caused during a search (clause 84.2)

Using ECC, I understand that if the *Supervisor* asks the *Contractor* to search for a Defect and no Defect is found, then a compensation event arises. Conversely, if a Defect is found, the *Contractor* remedies this at its own cost (depending on the main Option chosen). What happens when the *Contractor* damages a part of the *works* and essentially creates a Defect while carrying out the search? Is the *Contractor* entitled to recover the remedial costs as a compensation event or can the *Contractor* argue that creating a Defect was a risk of carrying out the test?

The answer is going to depend on the circumstances. If the test or inspection made the damage inevitable, it should be paid for as part of the compensation event because it was not accidental damage. So, for example, if the *Contractor* had to break out the newly constructed asset to test or inspect something underneath, then breaking it out and reinstating it is all part of the Defined Cost for carrying out the test or inspection.

On the other hand, if the damage was accidental in the sense that it could have been avoided but was not, then the *Contractor* is required to insure against such damage – see the first row on the Insurance Table in clause 84.2. In that case the cost will not form part of the Defined Cost – see the first bullet of item 7 in the SCC and the SSCC. The *Contractor* will have to look to its insurer to cover such costs and, if they are below the policy excess, that is the *Contractor*'s risk.

137. Accepting insurance policies (clause 85)

I am a *Project Manager* on an ECC3 Option C contract. I want to ensure that I do everything correctly but am concerned about by liability if I 'accept' a submitted insurance policy by the *Contractor*. Clause 85.1 requires that I 'accept' the policy but I am no expert in the insurance field. Where do I stand with this?

OK, the easy things to check for are that the amounts are compliant with the Insurance Table in Contract Data part one. You can also check that the certificates are signed by both the *Contractor*'s insurer and insurance broker (clause 85.1). However, reference to clause 14.1 should put your mind at rest as your 'acceptance' does not change liability. Finally, it is recommended that communication is undertaken in accordance with the 'How to … use the ECC Communication forms' guidance published in April 2013. This will ensure a clear audit trail and record.

138. Failing to provide insurance certificates (clause 85)

What happens if the *Contractor* does not provide insurance certificates as required by the contract/what is the remedy?

The contract requires (clause 85.1) before the *starting date* and on renewal of the insurance policy, the *Contractor* to submit certificates to the *Project Manager* for acceptance.

If they do not then the *Employer* may take out these insurances and the money is retained (clause 86.1).

This works the other way – the *Project Manager* is required to submit the same for risks that the *Employer* is required to insure against. If they do not then the *Contractor* can take out the insurance and the cost is paid by the *Employer*.

 **Some further thoughts and top tips**

- Make sure the Insurance Tables and amounts are appropriately completed at tender stage.
- Seek professional advice if at all unsure.
- Ensure that insurance details are provided to both Parties. Make a note of the renewal dates.
- Adding additional *Employer*'s risks is an excellent way of allocating risk fairly and obtaining better value for money.

Chapter 9
Termination

139. Is there a standard ECC termination certificate? (clause 90)

Q Is there a standard ECC termination certificate template available?

A NEC has not produced a termination certificate template as those produced are only for those most commonly used actions under the contract. Termination rarely happens and if it does we strongly recommend that you take advice to ensure that the contract requirements are followed. It is not something that should be done lightly.

The request for a certificate is issued by one of the Parties. The certificate will be issued by the *Project Manager* if he or she is satisfied that the reason applies and it complies with the contract. The certificate does not need to be any particular format; it just needs to comply with clauses 90.1, 13.1 and 13.6.

140. Is there a standard TSSC termination certificate? (clause 90.1)

We are using the TSSC and are looking to terminate in accordance with the provisions of clause 90.1 (reasons 2 and 3). The clause suggests that we have to issue a termination certificate. Can you advise whether there is a standard form for this, or whether it is anticipated that this will be a formal letter headed 'termination certificate'?

There is no standard form for this. The communication must be 'in writing' (see clause 13.1) and must make it clear what it is.

Please note that this is a two-part process. First the *Employer* needs to notify the *Contractor* that they wish to terminate, and then they issue a certificate if the reason given complies with the contract. You must also bear in mind with reasons 2 and 3 you must first give a notification of the default and then allow the *Contractor* 2 weeks to stop defaulting before you can notify your intention to terminate.

141. Can you terminate due to poor performance? (clause 91)

We are looking to terminate an ECC Option C contract due to a number of reasons; the main ones being poor performance in relation to programme and quality. How do we put this into effect?

Termination in the ECC terminates the *Contractor*'s employment under the contract, rather than terminating the contract itself (see clause 90.1). The actual reference is that you are terminating 'the *Contractor*'s obligation to Provide the Works'. The procedures for termination and the amounts to be paid on termination are set out in clauses 90 to 93 of the contract. Note the procedures and amounts will depend on who is terminating and for what reason – see the Termination Table in clause 90.2. Assuming you are the *Employer* here, it is essential that you provide a valid reason for termination (see clause 91) and then scrupulously follow the procedures set out in the contract. We suspect that clauses 91.2 or 91.3 would be most relevant to you, but in that case you need to notify the *Contractor* of the default and then give them 4 weeks to correct it, before terminating.

We are unable to advise you whether or not the particular 'poor performance' would meet the requirements of being 'substantial' required by these clauses, because it is in fact sensitive and will be a matter of law. You should therefore take legal advice on that point before proceeding.

142. Can we use Reason 21 termination due to minor earthquake? (clause 91.7)

On an ECC contract, if we suffer a minor earthquake during construction of the *works*, can we as *Contractor* terminate under clause 91.7 (R21)?

The simple answer is no, this right is only given to the *Employer* as stated in the first line of clause 91.7. In any case, there are a few hurdles to clear for this reason to be valid – one of the first two bullets in clause 91.7 must be satisfied along with both of the last two bullets. The operative word in the first two bullets is 'stops' not makes more difficult, delays or disrupts – the event that has occurred has to have stopped the *Contractor* completing the *works* (basically in its tracks, first bullet) or completing the *works* by the date shown on the Accepted Programme and be forecast to delay Completion by more than 13 weeks (second bullet). The clause goes on to test that neither Party could have prevented the event and finally that an experienced contractor (not the '*Contractor*') would have judged it at the Contract Date to have such a small chance of occurring that it would have been unreasonable for them to have allowed for it. When broken down, you can see there are actually quite a few hurdles to clear in clause 91.7 before the *Employer* could exercise the right to terminate.

143. Can you terminate for convenience? (clause 93.2)

When using ECC Option C, could you confirm whether clause 93.2 A4 applies in the case that the *Employer* decides to terminate after the 13-week time period stated in clause 91.6 (R20)? In other words, can the *Contractor* claim additional compensation in case of termination of the contract by the *Employer* for convenience?

The contract does not use the term termination 'for convenience'. Instead, the *Employer* is able to terminate for 'any reason' – see the second sentence of clause 90.2. If the reason is not one of those listed in the contract then payment A4 is made, which is effectively a payment of the Fee for the work that has not been carried out. That is seen by many as a level of 'compensation' as you call it. Therefore, the *Employer* can terminate at will, but they will pay more for that privilege if the reason is not one listed in the contract.

As for R20, you must remember what has happened beforehand. The *Project Manager* has already instructed the *Contractor* to stop work for a reason that is not the *Contractor*'s 'fault', which, in itself will be a compensation event (see clause 60.1(4)), and so the *Contractor* will have been standing around for 13 weeks, and have been paid for doing so. As an observation, it will rarely be the case that the instruction will have been for neither Party's default. That effectively means that a third party intervention or some other unforeseeable event will have occurred. For example, if the *Employer* runs out of money, or decides it wants to do the *works* a different way or with a different *Contractor*, or decides it does not want the *works* at all, those will all be the *Employer*'s 'default', so R20 would not apply. Instead the *Contractor* can either do nothing and be paid for it (note the *Employer* cannot terminate for R19), or terminate itself. If it chooses the latter using R19 then the *Contractor* will be paid A4.

In summary, the *Employer* who wants to terminate for a reason not stated in the contract will be unable to avoid paying A4 by telling the *Contractor* to stop work and then 13 weeks later terminating for R20.

 **Some further thoughts and top tips**

- People will often refer to 'terminating the contract', whereas in fact in ECC they are 'terminating the *Contractor*'s obligation to Provide the Works'. There is a difference.
- Termination is a very specialist area and great care should be taken to ensure that the exact provisions of the contract are followed.

Chapter 10
The Options

144. W1 or W2? (Options W1 and W2)

I am completing the Contract Data for a new build schools project and we are using Option A of the ECC. My question is – which dispute resolution procedure do I use?

Assuming you are in the UK the work you describe will be covered by the United Kingdom Housing Grants, Construction and Regeneration Act (1996). This will mean that you will need to use Option W2 – this has been specifically written to comply with the Act. You will also need to use Y(UK)2.

145. W1 or W2, which do I use? (Options W1 and W2)

Q I am involved in the construction of a major drainage project in Hong Kong. I am a bit confused as to which dispute resolution procedure I use – is it Option W1 or W2?

A You could use either. Option W2 has been specifically drafted to comply with the United Kingdom Housing Grants, Construction and Regeneration Act (1996). The main difference between Options W1 and W2 is that Option W1 contains time limits for the notification of a dispute. Many users prefer this option as you know where you stand in many respects.

146. Subcontracting and inflation (Option X1)

We are the *Employer* in an ECC Option C contract incorporating secondary Option X1 price adjustment for inflation. The *Contractor* has placed its subcontracts under its own bespoke form of subcontract, which does not flow down from the NEC contract we have with the *Contractor*. All of the subcontracts let are fixed-price lump sum plus any variations, and there is no mention of any escalation applying.

The *Contractor* is not paying any escalation amounts within its interim or final payments to its Subcontractors. The Defined Cost of Subcontractors is therefore exclusive of any escalation amounts. The question is, when assessing the *Contractor*'s price adjustment factor amounts under X1, should the amounts for these Subcontractors not be included within the Price for Work Done to Date?

They should be included, because the Price for Work Done to Date includes the costs of these Subcontractors – see clauses 11.2(29) and 11.2(23). Of course the answer you come up with for the price adjustment is not added to the Price for Work Done to Date, instead it is added to the total of the Prices (i.e. the 'target') – see clause X1.5.

Option X1 is just a convenient way of calculating an assessment of what the inflation may be, but there are all sorts of ways that the actual inflation can vary from that assessment. However, the risks, or rewards, of those variations are shared between the Parties through the share mechanism.

In reality, just because there are no inflationary increases within a particular subcontract, it does not mean that the *Contractor* will not be paying for inflation. In that case it could well be (and probably is) the case that there is an allowance for inflation within the Subcontractor's prices.

147. Indices calculations (Option X1)

We have an Option B ECC contract with X1 price adjustment for inflation. I am really confused by clause X1.3 – why do we need to alter the compensation event assessment to the base date?

At first read this is slightly confusing; however, the logic here is that each compensation event is adjusted back to the base date, i.e. at tender. Clause X1.2 then adjusts the amount due by the indices calculation. If compensation events were not adjusted in this way then the inflation would be double counted.

148. Adjusting for the price adjustment (clause X1.5)

We are using X1 on our ECC Option C contract. If I am reading X1.5 correctly it seems to be calculating the inflationary amount based on cost not the Prices as you would under Options A or B. Am I correct in this interpretation and what is the logic behind this?

Yes, you are correct. The calculation under Options C and D uses the Price for Work Done to Date. Under these Options this is Defined Cost (Subcontractors and for other work – the Schedule of Cost Components). Although it is almost impossible to assess correctly the actual amount of inflation that has actually been incurred on a project, this calculation attempts to be more realistic and derive the inflationary amount from the cost incurred as opposed to the Prices. In turn, this derived assessment is added to the Prices (clause X1.5).

149. I think I've found a drafting error! (Option X2)

Q I am completing the Contract Data on our ECC project and note there is no detail for Option X2 – is this a drafting error in the contract?

A Drafting error? No such thing! Option X2 is simply in or out so no additional information needs to be completed with Contract Data part one. The opening paragraph of Contract Data part one will determine whether this is used or not.

150. Is this a change in the law? (Option X2)

With reference to the recent changes in national insurance in the UK, the *Contractor* has notified a compensation event under Option X2. I would appreciate a commentary on admissibility in the light of the facts that the relevant statutes have been on the books from 2008 and 2010 and, hence, were known of at the time of tender. Furthermore, the details of the budget provided not only for an increase in overall rate but also in threshold – effectively providing a reduction for employees earnings under approximately £20 000 per annum.

With regard to Option X2 the position is clear – it is a compensation event if the law changes after the Contract Date. It is irrelevant whether or not that change of law had been announced many months or years before. So the question is when did the law change – not when was the change announced. That means that the *Contractor* is entitled not to allow for the effect of any change in the law that has been announced but not yet enacted. On the other hand, if the law itself had changed before the Contract Date but was not to take effect until a later date, then that will not be a compensation event. We do not know the exact arrangements for bringing in these changed national insurance contributions, so you will need to get expert advice on that. However, if the effect of this law change is to reduce employers' payments, then that needs to be taken into account in the valuing of the compensation event, i.e. reduce it (see the final sentence of clause X2.1). But note here we are referring only to the *Employer*'s contributions to national insurance, not the employee's.

151. Applying X2 (Option X2)

Q We are having a debate with our *Project Manager* on our ECC Option A contract. Option X2 was included but the *Project Manager* is saying that we should have predicted changes to legislation prior to the Contract Date and they should have been included in the tendered total of the Prices. Is he correct?

A No, he is not. Clause X2.1 clearly states that a change in law is a compensation event if it occurs after the Contract Date. Whether you could predict this occurring or even categorically knew it would occur is not relevant here. If a change in law occurs after the Contract Date then it is a clear compensation event. This could either increase or reduce the Prices.

152. Paying in different currencies (Option X3)

We are embarking on a processing plant in Yorkshire and we need to pay partly in Swiss Francs, Euros and Sterling UK. I appreciate that Option X3 is most appropriate but I am unsure how this works – please can you explain how I complete?

Yes, it is quite straightforward. You need to define the currency of the contract – presumably this will be Sterling UK. You will then need to define which items and activities will be paid in other currencies, what the other currency is and also the total maximum payment in that currency. It may be prudent to include an assessment for potential compensation events as this amount cannot be exceeded.

This final aspect to complete is the date when and by whom the indices are published. This will then allow the *Contractor* to assess this as part of its tender.

153. Parent company guarantee (Option X4)

We need to include a parent company guarantee on our ECC project but cannot find how to complete this in the Contract Data – please can you advise on the wording of the parent company guarantee?

We have not provided a sample parent company guarantee as this is something that is typically drafted on a bespoke basis by each client. You will need to seek specialist advice on this. The wording of this should then be included in the Works Information – this avoids any debate on the wording upon award. The parent company guarantee is then either provided at the Contract Date or within 4 weeks of the Contract Date. Note that this documentation is important to have in place – this is reinforced by virtue of clause 91.2 (R12). This states that the *Employer* may terminate the *Contractor*'s obligation to Provide the Works if this is not provided within the 4 weeks. 'May' means discretionary so the *Employer* has the option.

154. Is there a standard parent company guarantee form? (Option X4)

We are in discussion about a project using the PSC and there is a requirement for a parent company guarantee under Option X4. Is there a form produced by NEC for this?

There are no standard forms for such a guarantee as each organisation tends to use its own version. The version of the guarantee required by your potential *Employer* should be set out in the Scope, see the wording of clause X4.1 – 'in the form set out in the Scope'. If there is no form in the Scope, you will need to agree the wording with the *Employer*.

155. Defining *sectional* Completion (Option X5)

We need to have a *sectional* Completion on a project we are procuring under Option A of the ECC. The project involves the building of a new office block (for which we need a sectional Completion) the *Contractor* will then undertake the office refurbishment once we have vacated. We are incurring cost for every day we are renting the existing offices so I both want to incentivise the *Contractor* to complete early and protect ourselves against the cost of delay if they are late.

My query is how we should draft this in the contract to meet our objectives.

This is straightforward to incorporate into the ECC. Secondary Option X5 should be selected and the description of the *section* carefully drafted. You need to describe Completion of the *section* and the Completion Date. You may wish to consider using Option X6 (bonus for early Completion) and Option X7 (delay damages) to provide the incentive of early Completion and protection against additional costs if the *Contractor* is late. Under UK law the damages need to be a genuine pre-estimate of loss.

Finally, it is advisable to follow the Works Information guidance issued by the NEC in April 2013 – this provides support in establishing precisely the definition of *sectional* Completion (section WI 410).

156. How is *sectional* Completion confirmed? (Option X5)

We are currently using an ECC contract on a project that contains *sectional* Completion, all as outlined within Secondary Option X5 of the Contract Data part one.

Could you please confirm how and in what form *sectional* Completion is to be confirmed?

You deal with *sectional* Completion in exactly the same way as Completion – see clause X5.1.

Completion is a defined term – see clause 11.2(2) and that definition applies to Completion of any *section* of the *works* if Option X5 is used. The *Project Manager* decides the date when Completion (as defined in the contract) of any *section* occurs and certifies it within one week of that date – see clause 30.2. Certificates have to comply with clauses 13.1 and 13.6.

157. *Sectional* Completion – effects of delay (Option X5)

We are a *Contractor* under an Option C contract and the contract contains X5 and X7. The delay damages for the relevant section are £9000 per day. We planned to meet the *sectional* Completion Date some 7 days early but have been delayed by another *Contractor* (employed by the *Employer*) not providing us with information that we stated on the Accepted Programme. The *Project Manager* is stating that this float is 'shared' and as such he can use it. Please can you advise where we stand with this and the appropriate clauses?

This is exactly the same situation as if you planned to complete the whole of the *works* early and then you were delayed – this would be a compensation event. Clause X5.1 confirms that each reference to Completion (unless stated as the whole of the *works*) applies equally for the whole of the *works* or any *section*. This is a compensation event under clause 60.1(5). Clause 63.3 then confirms how this is assessed. If you planned to achieve Completion earlier than the *sectional* Completion Date and you were delayed by Others, then by whatever amount this is added to the *sectional* Completion Date.

158. Defects date running from Completion of the whole of the *works* (Option X5)

I attended an NEC training course and I am sure that it was mentioned that you do not have different defects liability periods running from *sectional* Completion. However, I cannot find where this is mentioned in the contract. Please can you advise?

That is correct – we do not have multiple *defects dates* running from each sectional Completion. In Contract Data part one it states that the *defects date* is a number of weeks after Completion of the whole of the *works* (typically 52 weeks).

159. How to calculate the bonus (Option X6)

We are a major supermarket client and are about to use the ECC contract for the first time. I note the ability to insert a bonus for early Completion, which would fit with our objectives as a client.

Is there a defined method of calculating the bonus for early completion or can it just be a figure that we come up with?

Unlike delay damages that need, under UK law, to be a genuine pre-estimate of loss – the bonus for early Completion (Option X6) can just be any figure you wish. Obviously, for internal probity/governance this will need to align closely with the benefit you get from early Completion.

Related to this is the need to have a clear definition of Completion. The default is that the *Contractor* has corrected Defects, which prevent the *Employer* using the *works* and Others from doing their work. It is recommended that you follow the guidance produced by the NEC on how to write the Works Information for the ECC (published in April 2013). This allows you in section WI 405 to define precisely what you expect at Completion. This may include operational and maintenance manuals, refrigeration units being tested, as-built drawings being handed over and user training on the new facility, for example.

160. Applying delay damages (Option X7)

I am a *Project Manager* on an ECC Option B contract. The *Contractor* is 20 days late on Site and we have delay damages of £3000 per day. They have been extremely effective on the project and we do not want to take the damages off them. Am I able, as *Project Manager*, not to deduct the damages?

As a *Project Manager* under the ECC you are obliged to act as stated in the contract (clause 10.1). This covers everything and, unfortunately, you cannot select which clauses to follow and which clauses not to follow. If delay damages are included and the *Contractor* achieves Completion after the Completion Date then delays damages are paid by the *Contractor* at the amount stated in the Contract Data multiplied by the number of days.

The only thing you can do is recommend to the *Employer* that damages are not taken. Under clause 12.3 if the Parties (that is the *Employer* and *Contractor*) agree then it can be put in writing and signed that delay damages are not deducted but this is not something, as *Project Manager*, that you can decide yourself.

161. Dealing with late Completion (Option X7)

Under the NEC3 ECC Option A contract, how would the *Project Manager* deal with a *Contractor* finishing late when no compensation events are apparent? Under the JCT contracts, the Contract Administrator would issue a Certificate of Non-Completion. Is this the case with the ECC also?

The ECC mandates the Parties to notify and deal with compensation events as they are known about. The processes and time scales are set out in clauses 61 to 65. And, as you seem to realise, the only way that the Completion Date can be extended to a later date is through that compensation event procedure.

We will assume that you have included the delay damages Option X7 in your contract. There is no such thing as a Certificate of Non-Completion in the ECC contract, because none is needed. That is because, unlike the JCT forms, it is the *Project Manager* that deducts delay damages as part of his assessment of the amount due to the *Contractor* – see clauses X7.1, 50.1, third bullet of 50.2 and 51.1. So, all that is needed is that, in the certificates issued after the Completion Date, the *Project Manager* deducts the delay damages accordingly.

We would also recommend that the *Project Manager* notifies an early warning, calls a risk reduction meeting (if this has not already happened) and tries to set about solving the problem of delay, if at all possible. Completion is a defined term – see clause 11.2(2). The *Project Manager* decides when Completion is achieved and certifies that date – see clause 30.2. In addition, the *Project Manager* must decide when take over has occurred and certify that as well – see clause 35.

162. Delay damages and take over (Option X7)

We are a *Contractor* on an Option D project. The delay damages are £90 000 per day. We were 3 days in delay when the *Employer* started to take over and began using part of the *works*. How does this affect the delay damages and how should this process be recorded and managed?

For the first 3 days the full £90 000 per day will clearly apply (see clause X7.1). The *Employer* is entitled to take over any part of the *works* before Completion has been certified. The *Project Manager* certifies this within one week of the date. If the *Employer* takes over before both Completion and Completion Date then this would be a compensation event. This is not the case on your project as you have gone past the Completion Date.

The bad news is that the *Employer* taking over is not a compensation event. However, the good news (if we may call it that) is that the delay damages are reduced by the proportional benefit obtained by the *Employer* in taking over. This is assessed by the *Project Manager* and typically may represent a significant reduction in the damages as most employers take over those parts of the *works* that offer most benefit.

163. Delay damages and gain share (Option X7)

We have an ECC Option C contract. The *Contractor* has achieved Completion after the Completion Date. The delay damages are £300 per day but the *Contractor* has also achieved a gain share. How do we calculate this – do we deduct damages from cost first then calculate the gain share?

No, the calculation is based on the difference between the Prices and the Price for Work Done to Date (clause 53.2). The Price for Work Done to Date (clause 11.2(29)) under Option C is based on Defined Cost (clause 11.2(23)). The amount due (clause 50.2) is a mixture of the Price for Work Done to Date, plus amounts to be paid to the *Contractor* (i.e. the share) less amounts to be paid by the *Contractor*, i.e. the delay damages. So, in answer to your question – work out the share first then deduct the damages after.

164. Applying partnering principles (Option X12)

We want to incentivise our *Contractor* to surpass the basic contract requirements on our ECC Option C project. We are also a real advocate of partnering and collaborative working. I presume that Option X12 is the appropriate Option to use – have the NEC produced any guidance on how this should be completed?

X12 sounds like it is not the most appropriate Option for you. This is for arrangements in which multiple partners are involved, e.g. on complex projects when the *Employer* wants perhaps his consultants, the main *Contractor* and key members of the supply chain to work together closely. An incentive schedule is completed and a Core Group is established to focus effort.

In circumstances when you just want to provide incentive to the *Contractor* (i.e. no other partner is involved) then Option X20 is the most appropriate. You will need to append an *incentive schedule* – this needs to define clearly Key Performance Indicators, how these will be measured and what the incentive mechanism is.

165. Getting performance bonds in place (Option X13)

As a local authority our standard tendering documents require that we include a performance bond. Where should the wording be included and how do we ensure that the bond is provided after contract award as we tend to find that *Contractors* are reluctant to provide once the tender has been won.

It is recommended that you follow the NEC guidance on how to draft the Works Information (published in April 2013). Section WI 1700 would be the appropriate place to include/append the wording. Specialist legal advice will need to be sought as to the wording itself.

In terms of making sure this is signed, you could get this signed on contract award. Failing that there is a strong provision in the contract – clause 91.2 (R12) states that if the *Contractor* does not provide a bond that the contract requires the *Employer* may terminate the *Contractor*'s obligation to Provide the Works. In accordance with clause X13.1 'may' is discretionary so it is up to the *Employer* whether they exercise this right.

166. Paying in advance (Option X14)

We are using X14 advanced payment to the *Contractor* on a nuclear project in the UK. We recognise that the *Contractor* has to make significant investment up front in order to meet the project timescales. Our question is, does the amount include VAT and what paperwork needs to be in place before the advanced payment is made?

You need initially to define the advanced amount – it should be made clear whether this includes VAT. If due on amounts to be certified later, then it would also be due on advanced payments. You will also need to define the instalments for repayment by the *Contractor*.

Finally, you will need to establish whether a bond is required or not (clause X14.2). The bond should normally be provided before the Contract Date. The advanced payment can be delayed until not later than 4 weeks after the *Employer* has received the bond.

167. Delayed advanced payment! (Option X14)

We are a *Contractor* on an Option A ECC project. We have just started on a high-value shopping centre complex. The contract includes an advanced payment but due to administration problems within the client's organisation the advanced payment is late in being made by a month. Where do we stand with this – are we entitled to interest and what about the disruption this has caused us?

Clause X14.2 states that any delay to the advanced payment is a compensation event and as such would be subject to the same assessment rules as any other compensation event, i.e. the cost and time both potentially need to be assessed. In addition, interest is due on the corrected amount (see clause 51.3).

168. Design liability (Option X15)

I am completing the Contract Data part one on our ECC Option B project. This is a civil engineering project with some tunnel design required by the *Contractor*. I understand that the inclusion of X15 is appropriate. I have included this in the opening paragraph but cannot find any further details to complete in the Contract Data – please can you advise? I understand it needs to be included, please can you also explain why?

There is nothing further to add and this is why no further details are required. You just simply state whether X15 is included or not in the opening paragraph of Contract Data part one.

Without this Option the liability must be in accordance with the Works Information. Under English law this would include a liability that the design is fit for purpose stated in the Works Information or reasonably implied.

X15 reduces this liability to 'reasonable skill and care' and reflects the insurance cover that a *Contractor* is able to obtain. The burden of proof, however, if a problem does occur, is with the *Contractor* to show they did indeed exercise reasonable skill and care not with the *Employer* to prove they did not.

169. Reasonable skill and care (Option X15)

 We are a *Contractor* delivering a complex prototype project in the nuclear sector (ECC Option C). Part of the *works* is unique and to the best of the knowledge of both ourselves and the *Project Manager* has never been undertaken before – Option X15 applies. In one aspect of the design both ourselves and the *Project Manager* agree that reasonable skill and care has been exercised by us in the design yet a Defect has still occurred. Our question relates to how this is dealt with – is this a compensation event?

 It is positive that the *Project Manager* is in agreement with you on this matter. In short, yes this is a compensation event under clause X15.2. You are still obliged to correct the Defect but in so doing this will be a compensation event.

170. Subcontractor retention (Option X16)

The *Project Manager*, that we are dealing with on an ECC option C contract is not including the amounts of retention we have deducted from payments to Subcontractors in his assessments of the Price of Work Done to Date as he says we are confusing the term 'retention' when phrased in the main contract clause 11.2(23) as having some connection with 'retention' in the subcontract. The *Project Manager* believes that these are entirely different concepts resulting from entirely different contracts. He states that Defined Cost at a main contract level should not have main contract retentions deducted as main contract retention is deducted from the Price of Work Done to Date if Option X16 is used. However, the *Project Manager* has also stated that any subcontract retentions should not be included in Defined Cost as these are an entirely different concept from the main contract retentions, and says that we should only be paid the 'amount of payments due to Subcontractors'. What do you think?

We believe the *Project Manager* is incorrect; all that you need to do really is to read and understand clause 11.2(23). The first part of the first bullet makes it clear that we are referring here to 'payments due to Subcontractors for work which is subcontracted', and the deductions listed subsequently in the sub-bullet points can only refer to deductions made from those payments.

If you look at the rest of the list, many of them can only apply to the payments you are making to the Subcontractor as they are clearly passing on things that you have had deducted, for example the second sub-bullet, or you have supplied to the Subcontractor, for example the last sub-bullet.

The only way that the *Employer* can deduct retention from you is if it uses Option X16. It cannot introduce it through the back door in this way. The ECC guidance notes make it clear that this is referring to deductions from the Subcontractor's payment, not the *Contractor*'s. It says that they are 'normally' deducted, but, in the case of the first one that will only be the case if the *Employer* uses Option X16. Just because they have not done what may be 'normal' does not change the meaning of the clause.

171. Retention percentage (Option X16)

We are a *Contractor* on an Option B ECC project. Contract Data part one states on the initial page that Option X16 will apply; however, the *retention percentage* was not completed by the *Project Manager* at tender stage. The *Project Manager* is now stating that normally this is 5% or 3% and, giving us the benefit of the doubt, it should be 3%. Where do we stand on this?

There is no 'normal' percentage as such. The guidance in the Contract Data warns: 'Completion of the data in full, according to the Options chosen, is essential to create a complete contract.' If not completed in full you simply have an incomplete contract.

You cannot just fill in the rest of the Contract Data with supposed norms/industry practice. If legal advice was sought then perhaps it would be found most favourable to yourselves as *Contractor* as we have one aspect of the contract mentioning retention and another section not. Perhaps the best way forward is to agree some sensible changes to the Contract Data to correct the drafting error – this can only be undertaken by the Parties (i.e. *Employer* and *Contractor* – appropriate representatives). It also needs to be in writing and signed by the Parties (clause 12.3).

172. Exceeding the retention amount (Option X16)

We have an ECC Option A contract – this is for the refurbishment of an existing swimming pool in a secondary school. For a variety of reasons the project did not go well. The re-lining of the pool was ultimately unsuccessful. Completion was certified by the *Project Manager* some 3 months ago and the *defects date* is 52 weeks after Completion. The original *Contractor* is just refusing to come back to rectify notified Defects. Our problem now is that the value of the Defects exceeds the 1.5% retention that is currently being held. How do we get this money back from the original *Contractor*?

OK, it is good that all the correct procedures appear to be being followed – Completion certificate and Defects being notified. It is important to maintain the notification of Defects (as per clause 42.2) even if the *Contractor* is not coming back at all. The contract states that (under clause 45.1) if the *Contractor* does not rectify the Defect within the *defect correction period* then the *Project Manager* assesses the cost to the *Employer* of having the Defect corrected by other people. The *Contractor* pays this amount. The contract does not say that you have to have the Defect corrected so you could just retain the money – presumably for a leaking pool the remedial work needs to be undertaken!

In terms of getting your money back, you need to assess the amount due under clause 50.2. This is a mixture of Price for Work Done to Date plus other amounts to be paid to the *Contractor* less amounts to be paid by or retained from the *Contractor*. This needs to be reassessed each month. If the balance of monies is now such that the *Contractor* owes you money then you need to raise an invoice. It is worth noting that clause 51.2 is written in a 'neutral' way in that failure of the *Contractor* to pay an amount they owe you would accrue interest. You will need to get your finance department to raise an invoice – if the *Contractor* fails to pay this then the normal debt recovery procedures should be followed.

173. Deciding the *retention-free amount* (Option X16)

In using ECC with Option X16 retention, can you advise on how the retention-free amount is calculated?

The term '*retention-free amount*' is in italics. That means it is an identified term (see clause 11.1) and the amount is therefore identified in the Contract Data. So, the amount will be decided by the *Employer* at tender stage and inserted in the Contract Data part one, under the optional statements for Option X16. It is entirely up to the *Employer* how much this amount is and how it is calculated.

174. Release of retention (Option X16)

We have an ECC contract incorporating Options X5 (sectional Completion) and X16 (retention). There are four *sections* defined and the retention percentage is 3%, with nil *retention-free amount*. Three of the four *sections* are complete and in use by the *Employer*. At present, one half of the retention has not been released in respect of the three completed *sections*. Retention is therefore currently held at 3% on the full value. The *Contractor* is of the view that one half of the retention applicable to the three *sections* completed should be released. The reason the retention has not been released is that clause X16.2 makes reference to the amount retained being halved at Completion of the whole of the *works*.

The whole of the *works* will not be complete until the final phase is complete so, until then, the retention remains at 3%. Clause X5.1 by reference to 'unless stated as the whole of the *works*' apparently excludes the application of the retained sum from each section. We think the *Contractor*'s interpretation perhaps reflects provisions of other standard forms of contract, but what is your view?

It is completely irrelevant what any other contract says (or not) about retention – it is all about what your particular contract says and we cannot fault your interpretation. If you simply read the relevant provisions through with the *Contractor*, the logical conclusion is that retention is not released for each section, only at Completion of the whole of the *works*.

175. How to draft low performance damages (Option X17)

We are considering using Option X17 on an Option C ECC project. This is for the construction of a new energy centre on our healthcare estate.

Please can you advise how this should be drafted?

The purpose of low performance damages is to reimburse the *Employer* where the performance of the *works* in use fails to reach a specified level due to a design or other fault by the *Contractor*. The required output should be clearly specified in the Works Information. It is advisable to follow the structure proposed in the published Works Information guidance (April 2013) – section WI 1900.

You will also need to complete the low performance damages in the Contract Data. These should be a genuine pre-estimate of loss. Under English law anything that is over and above that is generally not considered to be enforceable.

The low performance damages for say an energy centre could be graded based on certain levels of low performance.

The deductions for low performance damages are made in the Defects Certificate.

176. Limiting the liability (Option X18)

 Are delay damages included within the limitation of liability Option? i.e. are delay damages capped at a certain level?

 No, under clause X18.4 there are a number of 'excluded matters' (i.e. that are not subject to a limit of liability); these include (among others) delay damages and low performance damages.

177. End of liability date (Option X18)

We are a *Contractor* bidding for work that includes Option X18. We are trying to understand the significance of the *end of liability date*. This is completed at 10 years after Completion of the whole of the *works*. Is the end of all of our liabilities?

Yes, after this point the *Contractor* is not liable to the *Employer* for any matter, which would include a Defect, which is notified after this date.

178. Providing incentive (Option X20)

We have an ECC Option A project. We are a pharmaceutical client with tough targets to achieve on our project. We want to provide incentive if the *Contractor* betters the targets and also provide penalty if they fail to achieve them. X20 refers to an incentive schedule – presumably this is attached and covers both incentive and disincentive?

No, Option X20 is purely there to incentivise. The Incentive Schedule should be appended to Contract Data part one but should contain only incentives; these could incentivise the *Contractor* with say additional work on a framework or monetary payments. The Key Performance Indicators should be clearly defined, as should the method of calculating.

If you want to provide a penalty then the correct secondary Option is Option X17 – low performance damages. Under UK law these should be a genuine pre-estimate of loss.

179. Using X20 (Option X20)

Do the Key Performance Indicators in Option X20 have to provide incentive or can they be used purely for monitoring purposes with potential for future work?

The intent of Option X20 is to provide the *Contractor* with an incentive. Some *Employer*s just use Option X20 purely for monitoring purposes. The Key Performance Indicators are often aligned with corporate objectives. The Key Performance Indicators do not necessarily have to provide direct financial reward; they could offer potential for future work with an *Employer*. We suppose we have to respect the fact that the less direct a reward is, the less of a true incentive it actually provides.

180. Is a 'day' a day? (Option Y(UK)2)

We have recently started using a PSC and would like to know if 'days' used in time periods in the PSC means calendar days or working days?

The PSC only uses days, as opposed to weeks, to describe time periods in two places: the dispute resolution Option W2 and secondary Option Y(UK)2. Both of these are only used for work subject to specific UK legislation, as set out in the preamble at the beginning of both Options, and in order to comply with that legislation.

In both cases clauses within the Options clearly describe exactly what is meant by 'days' in the option – see clause W2.1(2) for Option W2 and clause Y2.1(2) for Option Y(UK)2. Although the wording is slightly different, the meanings are the same.

The only other time you may come across the word 'days' is if the *Employer* has used it when describing time periods in the Contract Data or the Scope. The meaning will depend on exactly what the *Employer* has written. If the *Employer* has just used the word 'days' without any other descriptor or qualifier, then it will mean exactly what is says: days. Saturday and Sunday are days, as are bank holidays, so they all count.

181. The purpose of Y(UK)3 (Option Y(UK)3)

Please can you explain the purpose of Y(UK)3?

This is only applicable on projects in England, Wales or Northern Ireland. This secondary Option ensures that only the terms set out in Contract Data part one can be enforced by those persons, or class of persons, that are named. It basically ensures that third party rights cannot be implied from any other terms of the contract or Works Information.

182. Preparing quotations for compensation events (clause 11.2(22))

We are using the ECC Option A contract. The contract specifically prevents the *Contractor* recovering the cost of preparing quotations for compensation events. What does preparing quotations mean? For example, would time spent obtaining quotations for compensation events from Subcontractors be caught?

The words mean what they say, and yes, obtaining quotations from Subcontractors is part of preparing a quotation.

183. Increase of resources by the *Contractor* (Option C)

In an ECC Option C contract, the *Contractor* is in 'pain' and will not recover the situation. The *Contractor* seems to be undertaking every administration activity possible on Site in what appears to be a move to make money on staff who would not normally operate on Site. Is it contractually acceptable for a *Contractor* to relocate staff that would normally be head office based (e.g. cost control clerk – bought ledger accounts) to Site to take advantage of the Working Areas Overhead percentage addition to the cost of people through the Schedule of Cost Components?

There also seems to be an increase in the attendance of the Site-based quantity surveying staff attendance from 1 to 3 days in order to do the same. There does not seem to be a proportional increase in performance or output from the surveyor related to our project and we believe this is an attempt to increase earnings.

The *Contractor* is entitled to organise its staff as it sees fit. And it is entitled to be paid for its people when they are in the Working Areas as long as they are Providing the Works, which, because of the definition in clause 11.2(13), would apply to these people. Of course, the more the *Contractor* increases the Price for Work Done to Date, the more pain it will end up paying. Therefore, this is a shared risk and, in reality, the *Employer* will claw back much of these Defined Costs at the end of the project.

184. Disallowed Cost (Option C)

Q We have an ECC Option C contract and are in a situation whereby a large metal sliding door has been delivered to Site in a damaged condition. Putting to one side the failings of the supply chain and all of the quality control procedures, the *Contractor* has then proceeded to install the door. The *Supervisor*, as duly obligated, has notified a Defect on becoming aware of the damage, i.e. after installation. Under the terms of the contract the cost of rectification of Defects before Completion sits with the *Employer*. Given the door has been delivered in a damaged condition, does this still apply in this instance? If yes, then if the door had been rejected at Site and returned straight back to the supplier for repairs/new door would the *Employer* still be liable for the costs of this? Note there is currently no direct sight of the supplier's terms and conditions but is this necessary given the condition of the goods supplied and the door still performs to its functional requirements.

A We are struggling to understand your suggestion that this problem sits with the *Employer*. Option C is, in fact, a shared risk contract, and the risk of Defects before Completion is one of those shared risks. This clearly seems to be a Defect and the *Contractor* is obliged to correct that Defect. The question is who pays what and when? Normally the *Employer* pays the cost, and therefore that increases the Price for Work Done to Date. However, the target stays the same (because this is not a compensation event) so the *Employer* recovers some of that back through the pain/gain mechanism on Completion. However, in this case, if the door was obviously damaged then the *Contractor* should have given the *Employer* an early warning of that under the fourth bullet of clause 16.1. Since they have not then one of the definitions of Disallowed Cost (clause 11.2(25)) is any cost incurred only because the *Contractor* did not give an early warning they should have. So, the additional cost of dealing with this problem now rather than before the door was fitted is a Disallowed Cost. So in theory, at least the cost of returning the door to the manufacturer before it was hung to get it sorted out would be paid. However, that is going to depend on the terms of the contract the *Contractor* has with his supplier, so you will need to see those. If the door was Defective on delivery then that will normally be down to the supplier to sort out. The *Contractor* has to keep those records and show them to the *Project Manager* under clause 52.

185. Productivity of resources (Option C)

Q Our contract is an ECC Option C project and our question relates to the SCC/Disallowed Cost. Item 11 of the SCC People 1 deals with people in the Working Areas. We have had it put to us by the *Project Manager* that our personnel have been (despite not actually providing the names of any individuals as evidence) taking too long for breaks; that walking to other areas of the Site, to cabins and/or to clocking stations is taking too long; that they have been standing around, and that they have left the workface early and been waiting at the clocking stations to clock out for periods of time. The *Project Manager* has stated that, in reference to section 11 above, 'time not worked' therefore is not payable under the SCC. Our interpretation of matters is that this is part of productivity or potential lack of it, which is catered for in the *Contractor*'s share. It is also our opinion that the interpretation of the above is that it gives a greater emphasis to the 'amounts paid by the *Contractor*', i.e. we get paid for the amounts we have paid. A secondary argument put to us is that the aforementioned bullet points are 'not used to Provide the Works' and so are not a Defined Cost anyway. Again our interpretation is that this is in fact resources used to Provide the Works as perceived lack of productivity is still Providing the Works, just not as effectively as we would wish. Your guidance would be appreciated on the above as it would appear that the *Project Manager* is attempting to remove substantial costs, which have been paid from the Price of Work Done to Date, to ensure that this does not even reach the calculation for that *Contractor*'s share and that therefore the *Employer* benefits in full from these deductions rather than sharing in the construction/productivity risk we believe this form of contract caters for.

A Our view is that the *Project Manager* is unable to do this. Dealing with the arguments, i.e. it is 'time not worked'. In fact item 11 is misquoted – this refers to 'people paid according to the time worked while they are in the Working Areas'. This merely conditions who this refers to, and undoubtedly the people involved fall within that category. It does not say that if they are paid despite not working it will not be allowed. And that is clear once you see the other matters that are paid for, such as travel time, absences for sickness and holidays, bonuses, incentives etc., which are all payments made that do not relate directly to the to the time worked. In fact, some of them, such as travel and holidays and sicknesses are specifically about 'time not worked' and yet they are still paid. In addition, you could successfully argue that even if they were not working then your paying them for not working was a 'special allowance'. Alternatively, you could successfully argue that the time taken to get to the canteen etc. was a payment made in relation to travel, as it does not specify what travel and to where. The simple fact is that if you paid these people in accordance with their terms and conditions, and normal construction industry practice, you are entitled to be paid for them as Defined Cost, and the *Project Manager* is, frankly, wrong to suggest otherwise. With regard to the second point, again that seems incorrect. These people were used to Provide the Works, and have been paid in accordance with their terms and conditions. And the costs therefore were incurred to Provide the Works. The only way that the *Project Manager* could try this would be to quote the penultimate bullet point of clause 11.2(25), concerning Disallowed Cost. However, even that would be a bad point because it is quite clear that you have to allow for 'reasonable availability and utilisation'. And this simply gets nowhere near jumping that hurdle. The *Project Manager* has clearly misunderstood how the Option C aspect works. As you correctly say, productivity issues such as these are a shared risk if productivity is low and shared reward if it is high. That is the whole point of using Option C.

186. Accrued cost in an application for payment (Option C)

I want some clarification of the inclusion of forecast or accrued costs in an application for payment. By this I mean costs that will be paid for in the following month, for which the cost is known but has not been paid yet; for example, the Site agent who is 100% allocated to the project. In theory, the *Contractor* knows the amount that the agent will be paid (in wages) but this cannot be substantiated yet. The same questions would be for the inclusion of estimated costs as well. I believe the answer is no to both, purely because at the point in time of the application, the costs cannot be substantiated. Is this correct?

We assume that you are referring to an ECC Options C, D or E contract. We consider that your opinion is correct. One of the definitions of Disallowed Cost is any cost that cannot be justified by the *Contractor*'s accounts and records. These sorts of costs cannot be justified by those records at the point when the *Project Manager* is making his assessment, so are disallowed. Of course, for the next assessment, when the payroll shows the payment, or the invoice for the estimated costs comes in, they will be allowed, even if they have not yet been paid (as long as they will be paid by the next assessment of course). This provision covers accruals made in the *Contractor*'s accounts for items that they have invoiced for (or other justification such as a Sub-contractor payment certificate), but that have not yet been paid.

187. The use of the Activity Schedule (Option A)

The tender documents were sent out and included an Activity Schedule and a suite of drawings. The successful tenderer included a BoQ in their tender reply, summed to a summary sheet based on the Activity Schedule. For some reason, the *Contractor* has priced some bill items that do not appear on the drawings. The *Contractor* is now asking for payment of the lump sum for the activity, which includes sums for items of work that have not been undertaken. We are acting as *Project Manager* and consider that if the *Contractor* has not done the work, it is not due the money. The *Contractor* argues that including the BoQ in the tender return was a mistake. I consider the bill to be a proxy for the *Contractor*'s duty to split the activities (groups of activities) into smaller packages to avoid conflicts of this very sort.

Option A is based on an Activity Schedule and that alone. It has no BoQ in it and there is no use for such an item anywhere. It should not have been included in the contract because it creates ambiguity, although it is not at all clear here about the status of either the BoQ or the Activity Schedule. If the BoQ was not included in the contract then it is, in any event, irrelevant.

Having said all of that, the contract is clear – it is the activities and prices in the Activity Schedule that matter and you cannot look to the BoQ. It is that, and only that, which states what the *Contractor* is to be paid once it carries out the works in Works Information. How and what it has allowed to get to that figure is irrelevant. Whether it has allowed too much or too little for whatever reason is its good or bad fortune. In order to demonstrate that, we would ask you to consider what would have happened if the *Contractor* had forgotten to include in its BoQ (and therefore its Activity Schedule) work that was in the Works Information? If it asked you to pay for them as an extra we suspect you would say 'no', which would be the correct answer. It is the same the other way round.

188. Items on an Activity Schedule not undertaken (Option A)

We are working on an ECC Option A contract. During the tender process and after inspection of the Site, the *Contractor* allowed for an element of hard breakout that was perceived as required in order to carry out the *works*. This amount was acceptable and included within the Prices with an annotation to describe the *works* as a description within a pricing document that was subsequently included in the Works Information. During the *works*, the hard breakout was not required as the assumed slab formation did not exist. Is the *Project Manager* correct in assuming these *works* can be omitted from the contract? And if so, please advise how this would be measured.

Option A is usually quite clear, it provides for a lump sum price to cover all of the work in the Works Information. The main way that the Prices will change is if there is a compensation event. It is not subject to re-measurement or any 'provisional' type items. This clarity seems to have disappeared in the way you have wrongly set the contract up.

Firstly, how you describe things in the Activity Schedule is irrelevant to what you are required to construct – see clause 20.1 and then clause 54.1. And secondly the Works Information does not state what you get paid – only the Activity Schedule does that – see clauses 50.2 and 11.2 (27). Further, the words in the Works Information only form Works Information if and when they comply with the definition in clause 11.2(19). Somehow you have managed to muddle the two different documents together, and it is difficult to suggest how this might be interpreted.

The answer may depend on what the words you put into the Works Information now say. They could be interpreted as meaning that if you do the hard breakout then you get paid, but if you do not you do not, which is in direct contradiction to the principles of Option A. They could equally be interpreted as being irrelevant because they do not comply with clause 11.2(19), so the amount of work would remain your risk (or reward). There is no clearcut answer here unfortunately, and the answer may depend on a legal analysis of the position.

Finally, it might be worth asking yourself what you meant when you put the words in there. And before you answer that, ask yourself whether the answer would be the same if you had found twice as much hard breakout as was allowed for. The same point should be raised if it was the *Project Manager* asking the question. To hazard a guess, you would probably each be using the other's arguments! If you had wanted to deal with the risk of this concrete fairly and openly at tender stage then you should have not allowed for it in the Prices and added the provision of a compensation event for finding it instead. It would then have been clear whether or not it was found would be at the *Employer*'s risk.

189. Additional management staff (Option C)

We have two questions listed below that we would appreciate an opinion on. We are using an ECC Option C contract:

1. The specification included in the Works Information identified a particular type of hinge but the *Contractor* qualified its offer at tender for this element, which was not identified or agreed. The *Employer* requires the originally specified hinge to be installed and the *Contractor* has now installed the hinge but has asked for a compensation event. This has been declined on the basis that the Works Information has not changed – is this correct?
2. The *Contractor*, without an early warning or agreement, brought additional management to staff in addition to the agreed management team. The application for their cost was declined as a Disallowed Cost on the basis that the resource was not agreed or necessary – is this correct?

The answer to the first part of your question may well depend on how and where the *Contractor* 'qualified' its tender. We say 'may' because the answer may end up with a legal (as opposed to contractual) analysis, which we cannot provide.

With regard to the second question the *Employer* has no right to 'pre-approve' resources the *Contractor* decides to bring to Site. Option C is a risk sharing contract. One of the risks you share is the *Contractor*'s incompetence; conversely, one of the rewards you share is the *Contractor*'s competence. It is for the *Contractor* to decide how best to manage the *works*, not the *Project Manager*. If these management staff were truly sitting doing nothing then they could be a Disallowed Cost – see the penultimate bullet of clause 11.2(25), but we doubt very much that you could show that to be the case. The *Project Manager* could request their removal and, a reasonable period after that, they would be a Disallowed Cost (see the wording of the same bullet). However, the *Contractor* would be right then to complain that the *Project Manager* is trying to manage the works on their behalf, and that could well, in itself, resolve into a dispute, so it is a dangerous path to tread. At the end of the day, the cost of these staff, like any benefits they bring, will be shared by the use of the share mechanism at the end of the contract.

190. Revising the Activity Schedule (Option A)

A *Contractor* submits his tender for an ECC Option A contract. The *Contractor* is not familiar with the payment mechanisms and only allows for a handful of lump sum activities. Within the Preliminaries section there is only one activity; the procurement and installation of sheet piling over the course of 2–3 months is one other. These two activities amount to approximately one-third of the Prices.

The *Contractor*'s tender is accepted and at the start of the project they are advised that they will only be paid on completion of each activity. Realising that this may create problems with cash flow, the *Contractor* submits a revised Activity Schedule with the activities broken down over the course of the programme. The *Employer* is concerned about the closure of one of the *Contractor*'s regional offices, although this is a major *Contractor* and a parent company guarantee is in place. The *Employer* does not believe that they are under any obligation to allow the Activity Schedule to be revised and therefore rejects the *Contractor*'s proposal.

Is the *Employer* in breach of clause 10.1 in that they are not acting in the spirit of mutual trust and co-operation?

Contractually, it is not the *Employer* that makes such decisions it is the *Project Manager*. The contract does not allow for the changing of the Activity Schedule, unless it is changed in accordance with the contract – see clause 11.2(20). The only way it can be changed in Option A is set out in clauses 54.2 and 63.12, neither of which covers these circumstances.

The spirit of mutual trust and co-operation cannot be used to change a contract the Parties have agreed to. So there is no contractual obligation to change this Activity Schedule.

Having given you the contractual answer, our view is that the sensible and practical answer is somewhat different. In our opinion it may not in the best long-term interests of the *Employer* for the *Contractor* to have a poor cash flow, with all of the possible problems that will cause to the *Contractor* and the supply chain during the lifetime of the project. Our practical view is that it may be more sensible for the *Employer* to enter into discussions with the *Contractor* to see if agreement can be reached on a more sensible Activity Schedule and then agree to change the contract in accordance with clause 12.3, but of course the Parties need to agree this and it must not fall foul of any procurement laws.

191. Interest paid to a Subcontractor (Option C)

The ECC Option C allows for interest to be paid on payments that are late. We are aware that the *Contractor* has made late payments to the Subcontractor and subsequently the Subcontractor has been paid interest. Does this interest payment then become due between the *Contractor* and *Employer*?

The answer is going to depend on the terms of the subcontract, not the (main) contract. That is because you are required to pay what the *Contractor* has actually paid his Subcontractor (clause 11.2(23)), but only to the extent that it is required by the terms of the subcontract (see the second bullet of clause 11.2(25)). If the sub-contract conditions require such interest to be paid, and if the *Contractor* has actually paid that interest to the Subcontractor then the answer is yes. However, they are two very big 'ifs', especially the second one. You should check to make sure that this amount has actually been paid, before paying it.

192. What are the *Contractor*'s share payment timings? (clause 53.3)

We are using ECC option C on a project. Can you explain the mechanism for payment when a project is in a 'pain' situation. When do any pain payments to the *Contractor* start for example? A worked example would be useful.

The timescales for making this payment are set out in clauses 53.3 and 53.4. A preliminary assessment is made when Completion is achieved, using estimates for the final Price for Work Done to Date and the total of the Prices (i.e. 'target'). The final assessment is made using the final figures. That is included with the final assessment, which is made 4 weeks after the *Supervisor* issues the Defects Certificate (see clause 50.1). These dates are the same whether the Parties (not just the *Contractor*) are in 'pain' or 'gain'.

The method of calculation is set out in clauses 53.1 and 53.2. A worked example is contained in the ECC guidance notes for those clauses where calculation (c) refers to a 'pain' situation.

193. Purpose of Z clauses (Option Z)

I hear lots of horror stories when it comes to Z clauses and recognise that a great deal of damage can be done through their use. Just for my own benefit, please can you explain the NEC's original intent behind Z clauses – if I know this I might be able to better defend their not being used in my own organisation.

This Option was intended when the *Employer* wishes to include *additional conditions of contract*.

The guidance notes give warning that Z clauses should only be used when absolutely necessary, such as changes required in a particular country.

We find that many Z clauses introduce legislation. This is unnecessary as clause 12.2 requires the *Contractor* to comply with the law. Also, many Z clauses introduce constraints – these should be included within the Works Information. So again, this is unnecessary.

The 'index' in the back of each contract within the NEC3 suite provides details on how the clauses are interlinked. This web/flow of the clauses can often be severed by Z clauses – if Z clauses are included it is always a good idea to check whether this flow has been impaired.

194. Deletion of clause 63.8 (Option Z)

We are a *Contractor* and are involved in an ECC project. There are a number of Z clauses on the project. One clause that has been deleted is clause 63.8, which explains how inconsistencies in the Works Information are dealt with.

Where do we stand on this if there are ambiguities or inconsistencies in the Works Information prepared by the *Employer* or indeed in our own?

This is where Z clauses can cause more problems than they try to solve! Clause 63.8 should, when left intact, work both ways. Ambiguities and inconsistencies are found most favourable to the Party that did not draft them, so inconsistencies and ambiguities in the Works Information prepared by the *Employer* are found most favourable to the *Contractor* and vice versa. If this clause is deleted then the contract does not resolve the issue. If it then ended up in the courts it is likely that the principle in clause 63.8 would apply as this is a recognised principle in common law – nothing is certain though.

195. Reducing the time bar (Option Z)

We are involved in a contract in which the 8 weeks has been deleted in clause 61.3 and replaced with 2 days. The time bar on the *Project Manager* has also been deleted under clause 62.6. Would these be enforceable?

Quite simply, if that is what has been signed then that is what will apply. You can probably guess how the contract is going to be run and how the *Project Manager* may behave if those are the amendments that have been made!

196. Deleting physical conditions and weather (Option Z)

We are involved in an ECC Option C contract in which clauses 60.1(12) and 60.1(13) have been deleted. Does this literally mean that there will never be a compensation event for physical conditions and weather?

Yes it does.

However, there are two instances when it perhaps does not accord with the perceptions of the drafter. One is that for every compensation event that occurs, the *Contractor* (under clause 63.6) is perfectly entitled to include risk for both time and cost. Is this really what the drafter of the Z clause intended – compensation events including more time and being more costly?

The second is that the cost will be incurred under Option C anyway dependent on the share ranges. In our opinion the deletion of these clauses is short sighted.

197. Complying with the law (Option Z)

I have previously been advised by lawyers that a Z clause requiring the *Contractor* to comply with the law is necessary. Have I been wrongly advised?

There is nothing that says you cannot do this, but it is hard to see the justification for it or how the Parties might suffer if such a clause were not present.

As an example, say there is a Z clause stating that *Contractor*'s staff shall not exceed highway speed limits. If one of the *Contractor*'s operatives speeds on the way to work, what are you going to do within the contract given that the *Contractor* has breached both the contract and the law?

The only justification for requiring compliance with the law as a contract term is that a breach of the law would also be a breach of contract. If some aspect of the law is so critical that you want to recover damages for the consequences of a failure to comply, that needs to be spelt out in the Z clause so people understand what is intended.

198. Stepping down Z clauses (Option Z)

Are Z clauses the correct location for 'step down' clauses from a main contract (e.g. IChemE) to a subcontract (e.g. the PSC)?

Yes, that would be the correct location. However, for the examples given, you would need to decide carefully which process-related clauses from the main contract would be appropriate for the professional services appointment.

You can potentially mix and match NEC contracts with others but you will need to take great care they fit together as intended. The simplest approach would be to use, say, the ECC as the head contract (which of course can be used for all process contracts).

Even after your best efforts when using a mix of contracts, you may still end up with one part of the supply chain using NEC language and processes (early warnings, compensation events, accepted programme) and another using entirely different language covering entirely different matters.

199. Mutual trust and co-operation (Option Z)

Surely some Z clauses are not compliant with ECC clause 10.1 requiring the Parties to act in a spirit of mutual trust and co-operation?

Some badly drafted or inappropriate Z clauses may indeed compromise the mutual trust and co-operation requirement, such that an inconsistency or ambiguity might arise. However, not all Z clauses will introduce inconsistency or ambiguity, nor will they compromise clause 10.1. Z clauses should only be included by an *Employer* if they are considered absolutely necessary, and then they should be well thought out and well drafted to avoid compromising standard NEC terms.

200. Spirit of the contract (Option Z)

What is the status of Z clauses that break the spirit of the contract? Case law is against a Party that badly amends a contract. From the *Contractor*'s point of view, is it a risk worth signing up to?

There is a point at which the spirit of the contract may indeed be broken by Z clauses. We would question why any *Employer* would choose a standard form of contract and then decimate it. Tenderers should carefully consider whether Z clauses are acceptable to them and act accordingly. Ambiguities and inconsistencies may well be resolved by those who created them – it is far better to sort such matters out before rather than after bidding.

201. Amending a core clause (Option Z)

Would it be correct to say that an amendment of a core clause is actually a Z clause?

That would be correct. Any amendment, addition or deletion of any core clause should be contained in a Z clause.

202. Replacing sentences (Option Z)

Q Instead of Z clauses, can you remove and replace core clause sentences?

A You can indeed, but what would you call this and how would you make sure such changes were properly incorporated into the contract? By pointing to the Z clauses in Contract Data part one and making sure all such clauses sit clearly under the heading 'Option Z: *Additional conditions of contract*', then there should be no doubt of their inclusion.

203. Protecting clients (Option Z)

For UK government contracts, *Project Managers* are prone to add Z clauses that are too protective for their clients. How do we balance such a tendency?

The best way for *Project Managers* to protect their clients is to use contracts with a balanced risk allocation, with each Party retaining the risks they can manage and allow for. *Project Managers* who tend to add Z clauses that result in an unbalanced risk allocation should be encouraged to focus instead on what is best for the project, which might help to balance such tendencies. The supply chain has a voice, and this should be used at the latest at tender stage to offer an alternative view when risk allocation appears to be unbalanced by Z clauses.

204. Using other main Option clauses? (Option Z)

Can you write a Z clause to amalgamate certain provisions from one NEC3 contract main option into another, for example using progress payment provisions from ECC Option B in ECC Option A? The reason would be to change the usual Option A requirement of requiring completion of a whole activity prior to payment.

Yes, it would be possible to do this. You would, though, need to be clear what benefit this would bring as presumably the intention would be to pay part-completed lump sums and you would have to make sure how this would objectively work in practice. Why not break down the lump sums into smaller activities in the Activity Schedule?

205. Valid Z clauses? (Option Z)

We regularly use Z clauses to deal with issues such as professional indemnity insurance requirements, assignment and collateral warranties. Are these valid uses for Z clauses or should we remove them?

It is important to determine what the perceived inadequacy is within, say, the ECC to decide whether indeed a Z clause is required. On the face of it, these are the sorts of matters that could well be appropriate Z clauses. There is, though, provision in the Contract Data part one to add in additional insurance requirements to that contained in the core clauses – professional indemnity insurance requirements should therefore be added here instead. If you consider that assignment is an essential provision between the two Parties then this should be added as a Z clause – but is it essential and what would happen (where a Party needed to assign) if such a clause was not included? A similar comment applies for collateral warranties: are these absolutely necessary and could the provisions of Y(UK)3 be used here instead, saving on drafting fees?

Some further thoughts and top tips

- Be clear on the project objectives and risks. Your selection of main and secondary Options should be focused on these.
- If using Options C, D, E or F ensure that a proactive open-book audit is undertaken at project start. This will ensure a clear understanding of cost and avoid differences of opinion later.
- Ensure the Options are properly completed. If not, we have an incomplete contract!
- Carefully consider the need for Z clauses. If they must be included check their impact by reviewing the Index (at the back of each NEC3 contract).
- Carefully consider the optional statements and include in the relevant section.

Chapter 11
Schedules of Cost Components

206. Time working within the Working Areas (SCC item 1)

Q With reference to the SSCC in the ECC, and in particular people who are directly employed by the *Contractor*, is it envisaged that time spent working away from the Working Areas would be classed as Defined Cost? If so, what would be considered as the proportion of time working within the Working Areas necessary to constitute the Working Areas being the 'normal place of working'?

A If a person's normal place of work is within the Working Areas, then their costs will be included in Defined Cost, even when they are not working in the Working Areas. So, if the Site Manager goes to head office once a month for a meeting, the Defined Cost for that is included. However, it is must be remembered that Site personnel's normal place of working will change regularly. For example, if Completion is achieved and the Site Manager moves to another Site, from then onwards his or her normal place of working will not be the Working Areas.

Other people may have no 'normal place of working'. For example, a quantity surveyor who looks after four contracts may visit each for a day a week and then spend a day in the head office. In that case he or she has no normal place of working at that time. If he or she is on your project one day a week then the Defined Cost, in simple terms, may be 20% of his or her total Defined Cost.

It is therefore simply not possible to make clear rules and percentages and it is going to come down to an objective view as to whether somebody has a normal place of working and, if they do, where that is.

207. Can we recover congestion charges? (SCC item 13(a))

We are currently working on an ECC Option C contract. Can you advise if the *Project Manager* is correct in his assessment of disallowing all costs in relation to London congestion charges? The *Project Manager* views these are not being incurred in order to Provide the Works.

Can these costs be claimed for under the SCC people item 13(a) – payments made in relation to people for travel?

'Provide the Works' is a defined term – see clause 11.2(13). It includes 'all ... incidental actions which this contract requires'. An 'incidental action' the contract requires is for the people to travel to work and that will include incurring congestion charges if they travel into or through the congestion zone.

This cost will therefore fall within item SCC item 13(a). Equally, it could fall within items 13(i) or 13(n). Whichever it falls within, it is incurred to Provide the Works, as is defined in the contract.

208. Meeting the requirements of the law (SCC item 13)

We have an *Employer* who is of the opinion that the CITB Levy we have to pay in regard to our people is not an acceptable part of the SCC and should be included within the Fee. Can you advise please?

We assume this query relates to the ECC. As far as we are aware, the law requires you to pay this levy. We believe it is governed by the Industrial Training Act 1982 and subsequent Levy Orders issued by the Government under that Act, but you will need to get legal advice to confirm that. In that case this is clearly a payment made in relation to people for meeting the requirements of the law and is therefore payable under SCC item 13(i).

209. Using estimated staff costs (SCC item 13)

We have recently started a contract using ECC Option C and have an issue regarding payment of people costs. The *Contractor* is submitting their direct staff costs for the agent, quantity surveyor and so on, on a pro rata allowance per week basis × actual staff cost. They are justifying this by stating that the quantity surveyor, for example, is allocated 2 days a week for the duration of the project and these are the costs we are getting submitted in the monthly applications for payment.

We believe this to be an incorrect method of valuing staff costs because they are not taking staff time from say timesheets of actual time spent in the Working Areas. On this particular project, the *Contractor* is charging us a standard allowance per week regardless of the number of days actually spent. Are we wrong in this?

The term used in the contract is Defined Costs and for Option C the definition is set out in clause 11.2(23). The last part of this requires that you use the SCC to value such cost. The preamble to SCC item 1 sets out what people are paid for and when. As this quantity surveyor is not based in the Working Areas it is the second bullet that applies in their case. And that requires that such people are paid only for the time that they are working in the Working Areas. The quantity surveyor cannot be paid for while he is working anywhere else, even if that work is on the work involved with your contract. That cost is assumed to be in the *Contractor*'s Fee – see the first sentence of clause 52.1. Although you make no mention of it, you will need to make sure that the costs are paid on the amounts paid as set out in SCC items 11, 12 and 13. You have the right to see and check such amounts, including looking at payroll records, under clauses 52.2 and 52.3.

210. Payment for pension shortfall (SCC item 13)

We have an ECC Option C contract. Would the *Employer* have an obligation to pay *Contractor* pension shortfall (incurred by other means)?

The simple answer is yes, but only the amount that relates to the time people spent in your Working Areas. So if the person was on your Site for a year, but in the pension scheme for 10 years, then you would pay only for your one year of catching up. How that would be calculated and whether it was lineal or not is something you will need to discuss with the *Contractor*.

211. Dealing with medical expenses (SCC item 13)

We are currently running a ECC Option C on Site (£9 million, 30 month programme). Our *Contractor* has put through a claim for a staff member medical expense against the contract for £11 000. It seems the original (private) hospital healthcare bill came to roughly £13 000 and the *Contractor*'s insurance company has settled to pay £2000 of the bill. The *Contractor* has looked to settle the remainder of the £11 000 bill by adding as a contract cost, which they are claiming in relation to people for medical aid under the SCC item 13(m). Is this acceptable?

We have no concerns that the cost was incurred and that the staff member is a full-time staff member of the project. We feel that if the *Contractor* had suitable healthcare insurance that covered the full cost, then we would have been happy that the premium was a contract cost. However, the excess should have been covered by the individual or the *Contractor* and not the *Employer*. This has resulted in £11 000 of unexpected cost to the contract, which was not allowed for.

Paying this will be a judgement call we suspect. Firstly, why did the medical insurance company only pay for a small amount of this? Was the rest considered to be too expensive or unnecessary? If that is the case then you could say that it does not comply with clause 52.1 and should not be paid. Next, what was this for? Was it to enable the person to be cured quickly of something that was going otherwise to keep him or her off work or prevent him or her working properly (until the National Health Service could eventually cure it) or was it purely cosmetic? The former is likely to be payable, the latter is not because the 'market rate' would be the National Health Service's cost, which would be nil – see again clause 52.1. Also, was it something that you would normally expect the National Health Service to carry out just as quickly and as well? In that case again clause 52.1 would say that this was not at the 'market rate', which was nil. And finally, what does the person's contract of employment say about all of this? In our experience, it is very unusual for the *Employer* (in this case the *Contractor* of course) to pick up any excess on medical insurance, or any other amounts that the insurer will not pay. Most *Employers* insist that the employee pays those amounts. If it is not an obligation for the *Contractor* to pay this then you should not be required to pay it. After all, the *Contractor* (not the employee) has to be shown to have paid this amount, and they have not done so. If all of these questions answer in the favour of the *Contractor* then we think it should be paid by you.

212. Hand tools not powered by compressed air (SCC item 44)

We are using the ECC. Is there a clear or definitive list as to what is considered a hand tool not powered by compressed air? We are encountering a number of arguments with contractors about this issue and would like to resolve it once and for all.

The simple answer is that there is no definitive list, and nor could there possibly ever be, because of the huge range of tools available for the many different activities carried out using the ECC, which changes as technology and the way things are done changes. This provision therefore needs to be interpreted in accordance with common sense and the law. Almost every machine, from an excavator to an aeroplane, is usually operated (or at least partly operated) by hand, so that cannot possibly be the criteria. And the fact that tools powered by compressed air are specifically excluded means that tools powered by other means are not necessarily excluded. Instead, the criteria means that the tool is held, in the sense of supported, by the hand.

So, to give an example, an electric circular saw or jigsaw is a hand tool, but a saw bench is not. An electric drill is a hand tool but a pillar drill supported on the floor or on a bench is not. Hydraulic shears held in the hand are a hand tool, but the machine that provides the hydraulic pressure is not. All of this also means that items such as hand saws, hammers, screwdrivers, chisels, trowels, shovels, picks, etc. are all hand tools. It also means that if the *Contractor* is breaking out concrete using a small electric breaker it is included in the Working Areas overheads percentage (and is therefore not paid for as Equipment), but if he is using a small compressed air breaker it is not included in the Working Areas overheads percentage (and therefore is paid for as Equipment). But in either case the generator (if not mains electric) or compressor used to power the tool is not part of the Working Areas overheads percentage and is therefore paid for as Equipment.

213. Do we get the Working Areas overheads on top of Subcontractors' costs? (SCC item 44)

We are on a project under the ECC Option C. Do Working Areas overheads apply to a Subcontractor who is employed on a place-only contract for reinforced concrete and rebar installation. The unit rate allowances cover items such as the provision of plant and or tools, labour, materials and supervision.

The answer will depend on what type of contract the *Contractor* has with its Subcontractor. If they have a contract to place the rebar and are paid based on how much they lay, they are defined as a 'Subcontractor' – see clause 11.2(17) – they have a contract to install part of the *works*. As such you do not use the SCC at all to decide how they are paid – see the first bullet point of the definition of Defined Cost in clause 11.2(23). In addition, payment to a Subcontractor is not included in any part of the SCC and therefore Working Areas overheads percentage is not added to it.

If the contract only requires that the organisation provides skilled labour that is paid for by the hour, and the *Contractor* tells them what to do, then that does not fall within the definition of a Subcontractor in clause 11.2(17). This is because they do not have a contract to install part of the *works*, just a contract to supply labour. In that case they are paid as part of the labour element in the SCC – see item 14. In that case the Working Areas overheads percentage is added, as clearly stated in item 44 of the SCC.

214. Is security in the Working Areas overheads percentage? (SCC item 44)

Item 44 of the SCC contains a description for the Working Areas overheads percentage that includes security (excluding accommodation). We have a contract in place in which the *Contractor* is seeking payment for the security staff under people item 11. The *Contractor* claims that all people are included in item 1 and then the Working Areas overheads percentage is applied to this. The *Contractor* also suggests item 44 only covers security items such as padlocks, fencing, etc. How should we deal with the *Project Manager*'s assessment of this?

On a similar subject, on an ECC Option C contract, we have described within the Works Information the security arrangements required for the *works*. Is there an entitlement for the *Contractor* to be paid additional anything above the Working Areas overheads percentage, for example within people item 1 of the SCC? For clarity, please could you provide examples of the items that we should expect to be included within the Working Areas overheads percentage relating to security.

The Working Areas overheads percentage in SCC item 44 covers 'the provision and use of equipment, supplies and services ... for ... security'. People carrying out that security cannot be described as being equipment supplies or security and they are therefore not covered by the Working Areas overheads percentage.

In the same way the people working with the telephone, surveying equipment, computers or hand tools (all included in the list) are also not included in the Working Areas overheads percentage. So the security guard is paid for as people. However the burglar alarm and security fencing is 'equipment for security', the padlocks are 'supplies for security' and the servicing of the alarm is 'services for security' so they are all included within the Working Areas overheads percentage and are not paid separately.

The answer is the same whether or not you have specifically required any particular security arrangements in the Works Information. The rules in ECC Option C as to what the *Contractor* gets paid are the same either way.

215. How is a security guard paid for? (SCC item 44)

In an ECC Option C contract, the SCC in item 44 states that provision of services for security is covered by the Working Areas overheads percentage. If the *Contractor* employs a security firm to undertake security work within the Working Areas, which are paid on an hourly subcontract basis, is the *Contractor* entitled to the security people cost component as covered by item 44 or is it covered by the Working Areas overheads percentage?

The Working Areas overheads percentage only covers 'equipment, supplies and services...' for the listed items in item 44 of the SCC. It therefore does not include security guards providing the security, any more than it covers setting out engineers using the surveying and setting out equipment, or the Site Agent using a computer. Those are all paid for as people in item 1 of the SCC. In your case it would seem that the guard would probably fall under item 14.

216. Working Areas overheads percentage on people (SCC item 44)

In an ECC Option C contract, item 44 of the SCC allows for the Working Areas overheads percentage to be applied to the Defined Cost of people. The percentage applied under item 44 relates to the SCC items 11, 12, 13 and 14. Currently, we only apply the Working Areas overheads percentage to the people cost. Following an independent audit, the auditors have determined that this should be applied to people and the associated expenses under items 11, 12, 13 and 14. For example, the auditors are saying that the Working Areas overheads percentage should be applied to a person and also his or her associated expenses for say travel and lodging allowance. Which is correct?

The auditors are correct. There seems to be no logic for not paying the Working Areas overheads percentage on these amounts, and we are surprised that the *Contractor* has not queried it. Items 11, 12, 13 and 14 are all people items in the SCC, because they come under SCC item 1, which has a heading of 'People', and because they all relate to payment made to, or in relation to, people. And item 44 makes it clear that the Working Areas overheads percentage gets added to the 'total of the people items 11, 12, 13 and 14' and that is exactly what you do. Travel and lodging allowance (among many others) all come under SCC item 13(a), (b) and (h), respectively, and are all people items. If the Working Areas overheads percentage was not to be paid on these items then item 44 would not have included a reference to item 13. However, it does because the Working Areas overheads percentage is to be paid on these.

217. What does 'services' mean in the SCC? (SCC item 44)

 In the ECC contract, what does 'services' in SCC item 44 cover?

 The term 'services' could be things like emptying the portaloo (sanitation), servicing the alarm (security) or repairing the copier.

218. Using the SCC Working Areas overheads percentage (SCC item 44)

Q In an ECC Option C contract, can you please advise when the Working Areas overheads percentage as stated in the Contract Data is applied and to what it is applied?

A In Option C you must use the SCC to calculate the Price for Work Done to Date – see clauses 50.2, 11.2(29) and 11.2(23) in that order. You also use the SCC to assess compensation events (see clause 63.1). In the latter case, however, you may use the SSCC if both Parties agree to it (see clause 63.15). However, that rarely happens because once you have the costs each month to calculate the Price for Work Done to Date there is little point in using a different method to calculate compensation events. The Working Areas overheads percentage gets added to the cost of people if you are using the SCC – see item 44 of the SCC. The people overhead percentage gets added to the cost of people if you are using the SSCC – see item 41 of the SSCC.

219. Damage to Equipment (SCC item 7)

We have a query on an ECC Option C contract concerning Disallowed Costs. We are questioning repairs to a *Contractor*'s camera, damaged on Site. Is this a valid Defined Cost, or not? We can see the hire is allowable under SCC item 22, but can see no item whereby the repair is an allowable cost.

You have rightly identified this camera as Equipment. The Insurance Table in clause 84.2 requires the *Contractor* to insure loss of or damage to Equipment. The first bullet of SCC item 7 makes it clear that you deduct from cost the cost of events for which the contract requires the *Contractor* to insure, and that includes this event. You do not therefore pay for this. The fact that the damage will be below the excess on the insurance policy is at the *Contractor*'s risk.

220. Damage to vehicles (SCC item 7)

We are using ECC Option C. The *Contractor* has included within Defined Cost the cost of repairing damage to vehicles that have been used by the *Contractor*'s employees over the past 2–3 years. The cost of a replacement hire vehicle has also been included. Such costs have been included within Defined Cost on the basis of SCC item 1 people part 13(n) provision of a vehicle. The vehicles in question are owned and repaired by another division of the Contractor.

The *Project Manager* considers that the cost of damage repair and hire falls under the SCC, section 7 insurance bullet point 1, and as such adequate insurance provision should have been provided under clause 84.2, row 3 of the listed table. The *Project Manager* has disallowed significant repair and/or hire costs on the basis of clause 52.1, which states that any *Contractor*'s costs that are not included in the Defined Cost are treated as included in the Fee. Does the cost of repairing damaged vehicles and the cost of replacement hire vehicles fall under 13(n) of the SCC as Defined Cost or should it be treated as included in the Fee?

These vehicles are not owned by the *Contractor* but by a separate legal entity, albeit one with the same parent company. As such, the *Contractor* would be liable to that separate entity for any damage that occurs to their property. Therefore, under the third row of the Insurance Table, the *Contractor* is required to insure against such liability. The costs of these accidents could well fall within SCC item 13(n); however, the first bullet of item 7 of the SCC makes it clear that such costs are, nevertheless, to be deducted because they are the cost of events that the contract requires the *Contractor* to insure against. The *Project Manager* is therefore correct not to allow such costs.

221. Can we recover the costs of transport for people? (SSCC item 1)

We are using the ECC and have some compensation events valued using the SSCC. An item we have claimed for under the Equipment heading is for a work van used to transport people to and from the Working Areas. The *Project Manager* has struck this cost out in our quotations. The published list in the Contract Data is the CECA Schedule of Dayworks. Under Item 3 Supplementary Charges within the General Notes it says at point 1 'Transport provided by contractors for operatives to, from, in and around the site to be charged at the appropriate Schedule rates'. Is the *Project Manager* correct not to include the cost of the van in our quotations, or can we recover this as a cost?

You cannot claim for this van as Equipment, because that only covers Equipment used in the Working Areas, not transporting people to those areas – see the first sentence of point 2 in the SSCC. It does not matter what the General Notes in the CECA Schedule says, because you only use the Schedule to get the rates for the Equipment that is to be paid for in accordance with the SSCC.

The cost of this, however, can be included as part of the cost of SSCC item 1 people. It is part of the people costs. If you look at the ECC guidance notes for this it confirms that the intention is that these people costs would include all the specific items listed in the full SCC. And that cost includes payments made in relation to people for the cost of travel (see item 13(a) of the SCC)).

222. Cost of people in the SSCC (SSCC item 1)

Under the SSCC in the ECC Option A there is a cost component for people in item 1. It is unclear what can be included in the cost of people and it is our view that it is made up of the cost to employ the person (item 11), i.e. salary of the person plus tax, national insurance and pension contributions, plus any taxable allowances, for example a car allowance. We do not believe the cost of people includes the cost of such items as travel, subsistence and lodging as we believe these are to be deemed included in the people overheads. Are our views in line with the intended application of the SSCC?

Also with respect to the taxable allowances, such as car allowances, is it included in SSCC item 11 even if it is not necessary to deliver the *works*? A literal and strict reading of SSCC item 11 would suggest that allowances should only be included if it is to meet the requirements of the law and pension.

SSCC item 41 tells you what the percentage for people overheads includes. It does not include those costs you have listed, so if they are a cost for employing the person, which they are, then they are amounts that are paid as part of the cost of employing the person. So you should include them.

With regard to the second part of your question we are not sure where you have found the term 'necessary to deliver the *works*', because that is not the test as to Defined Cost in Option A. If the person's Defined Cost changed as a result of the compensation event (which we assume that it did as you are assessing their cost) and the payment is part of their contract of employment, then the cost was 'incurred in order to Provide the Works', which is the test to be met – see the preamble to the SSCC. And we cannot see how a 'literal and strict' (or any other) reading of SSCC item 11 could make any difference. Just because it includes 'meeting the requirement of the law and pension provisions', does not mean that is all it includes, otherwise you would not allow for their wages or salaries would you?

223. Which SSCC Equipment rates do we use? (SSCC item 2)

We are currently engaged on an ECC Option A contract. At tender stage, in our Contract Data part two – data for SSCC, we submitted a percentage reduction for Equipment under the CECA Schedule of Dayworks. We also submitted a rate under 'other Equipment' for a gang consisting of a 7 t lorry, a mini-excavator with rock breaker and a transporting trailer. This gang was dedicated to this particular contract, which involved minor works at dozens of different locations.

We have since submitted the 'other Equipment' gang rate in quotations for compensation events but the *Project Manager* has assessed the compensation events at a lower value by breaking our gang down to its constituent parts and applying the CECA Schedule with our reduction. We can see where he is coming from but we contend that we submitted a gang rate in our tender, that gang was used for compensation event works and we are entitled to compensation event assessment at that rate. Please advise which rate should take precedence?

The 'other Equipment' is supposed to be used only for Equipment that is not within the CECA rates, which is why the entry in the Contract Data follows the CECA entries and therefore refers to only Equipment 'other' than can be found in the CECA rates. It is for specialist items of Equipment, such as piling rigs, tunnel boring machines and dredgers, which are not covered by the CECA rates. And that can be seen when you look at SSCC items 21 and 22. It is clear that item 21 is used for all Equipment for which a rate can be found in the published list, and item 22 is only used for Equipment 'which is not in the published list stated in the Contract Data'. Therefore, contractually, the *Project Manager* is correct in what he is doing.

224. Cost of Equipment standing (SSCC item 2)

We are using the ECC Option B and would be grateful for your thoughts on the following event. On the first day of a construction project the *Contractor* had to stand down an excavator, which lasted a matter of weeks. A compensation event has been notified with the Equipment priced at the rate listed in Contract Data part two to be recovered for the period of delay. It is our view that Equipment listed in Contract Data part two is meant for incidental works and is not necessarily intended for use when there may be significant delay.

We have subsequently suggested that the Equipment should be priced at market rates or equivalent hire cost as depreciation/fuel costs would be reduced as a result of the stand down. Secondly, if the *Contractor* had programmed to bring Equipment to Site but did not as a result of the stand down, we feel that as the Equipment was not within the Working Areas, it cannot be charged for. Is this correct?

The contract is clear that you use the rate in the Contract Data – see either items 21 or 22 in the SSCC depending on which applies. Terms like 'incidental works' are irrelevant because they are not used in this contract. If you thought the rate was too high (or the percentage was wrong) then the time to raise that was before you accepted the tender. Once you have a contract it is clear that is what you pay.

With regard to your second question, an initial inspection of the contract would suggest that you do not pay this cost because the Equipment was not in the Working Areas. However, it is not as simple as that. The compensation is based on a forecast of the Defined Cost, not the actual Defined Cost. That is because the last sentence of clause 63.1 makes it clear when you stop using actual Defined Cost and clause 61.1 makes it clear that is the date when you issued the instruction we are talking about. Once you use a forecast you do not change it based on what actually happened – see clause 65.2. So would a reasonable forecast be that the *Contractor* brought the Equipment to Site? Given that they will not be paid for it unless they do (and they are aware of the circulatory nature of that statement) then we suspect the answer is yes. If you were to allow for it we think you have the right to check to make sure that it was, indeed, standing and not being used somewhere else. In reality, there is no absolutely definitive answer to this one. We suspect that if the *Contractor* could show they were not using it, some *Adjudicators* would allow payment for it, although at what rate is another matter! Others would stick to the literal meaning of the contract and not pay for it.

225. Using the percentage for people overheads (SSCC item 41)

Our query relates to the ECC and Contract Data part two, specifically where Option A or B is used (although there are similar requirements elsewhere). In the generic ECC it states 'Data for the Shorter Schedule of Cost Components' and asks for 'The percentage for people overhead is ... %'.

We have read the contract and the guidance notes and are still unclear on this. Does this percentage reflect the value of all staff costs for this project/value of the tender, or the value of the staff costs for this project/value of the overheads of the project, or something else?

The SSCC in Options A and B is only used to assess compensation events. These are assessed based on the effect they are forecast to have on Defined Cost and then Fee is added (see clauses 63.3 and 11.2(8)). Defined Cost is defined in clause 11.2(22), and is calculated by using the SSCC. The Defined Cost of the people involved in compensation events is calculated based on what they actually cost to employ (see SSCC items 1 and 11). To that people component will be added the percentage for people overheads you state in the Contract Data, which is specifically meant to cover those items set out in SSCC item 41. You will not be paid those items because you will have been paid for them within this percentage (see the final sentence of the preamble to the SSCC). Therefore, it has nothing directly to do with either of the options you set out in your question.

226. Using the SSCC percentage for people overheads (SSCC item 41)

In an ECC Option C contract, can you please advise when the percentage for people overheads as stated in the Contract Data is applied and to what it is applied?

In Option C you only use the people overhead percentage if both parties agree to use the SSCC to assess a compensation event (see clause 63.15) and this overhead will then get added to the cost of people (see item 41 of the SSCC). The *Project Manager* may also make his or her own assessments using the SSCC.

227. Using the percentage for design overheads (SSCC item 6)

Please can you confirm what the percentage for design overheads in the ECC Option A is and how this is different to the Fee, as we need to quote these percentages in a tender?

As part of your tender you will need to complete Contract Data part two, where you can find reference to this item. With Option A these figures are only used to calculate Defined Cost, which in turn is only used to assess compensation events – see clauses 63.1 and 11.2(22) along with the SSCC. In Option A the only way (with minor exceptions) that you will get paid more or less than your tendered Prices is through compensation events. With regard to the Fee (not for design) you need to look at clause 11.2(8). You have to quote two fee percentages, a *subcontracted fee percentage* and a *direct fee percentage*. The former gets added to the Defined Cost work that you subcontract (see clause 11.2(17)), and the latter gets added to all other work. What that has to allow for is up to you. The SSCC sets out what you will get paid for when calculating Defined Cost; anything else you want to recover for compensation events has to be included in your Fee – see clause 52.1.

The way that you calculate the Defined Cost for design (if any) for compensation events is set out in item 6 of the SSCC. Basically in the Contract Data you quote an hourly rate for different grades of designers and a percentage that gets added for design overheads (it is not called a fee). It is entirely up to you what you want to do with these figures. If the hourly rate you quote for designers already includes all the overhead and profit you wish to recover then the percentage for design overheads can be zero. Otherwise you can make the figures whatever you want; the proviso is of course that if the *Employer* thinks they are too high it may not accept your tender.

 **Some further thoughts and top tips**

- The purpose of both the SCC and SSCC is to assist in the clarity of understanding Defined Cost. Both will need careful studying to aid the tender process and later contract management.

Chapter 12
Contract Data

228. What *services* are in the PSC Scope? (Contract Data)

Q A *Contractor* has been employed on the PSC to provide overall management of a programme of *services* to include whole lifecycle activities from competition to completion. My question is, does this include the services of an ECC *Supervisor* or on-Site health and safety auditor, or is this a role that the local authority would procure?

A The answer will depend entirely on what is written in the relevant PSC Scope. If it says the *Consultant* (the name used in the PSC) provides this *service*, then it is included. Otherwise the answer will not be clear and will depend on an interpretation of the words used in the Scope to define precisely what the *Consultant* does.

229. Who is Mr I Judge? (Contract Data)

Q We are about to use the ECC for the first time on a new schools project. We have never completed the Contract Data previously – could you direct us to any relevant guidance?

A Each of the contracts within the NEC3 suite contains accompanying guidance notes. The ECC guidance notes contain guidance on how to complete the Contract Data (Appendix 5). This is annotated in order that you can refer back to the relevant section in the guidance notes themselves. Do not copy it exactly as we have seen a real contract with Mr I Judge as the *Adjudicator*. This was meant to be a joke!

230. How do you describe the Affected Property? (Contract Data)

We are preparing tender documents using the TSC. When an Affected Property is made up of 10–15 buildings (which form an establishment) what can we refer to these as?

The first thing you need to decide is exactly what the Affected Property is. Is it 10 or 15, or any number in between? Once you know that you can describe it in any way that is sensible and informs the *Contractor* exactly what buildings he is required to provide services to. So you can show them on a drawing and refer to the drawing number in the Contract Data. Or list them in the Contract Data. Or say 'all the building at XYZ's site on the High Street, Anywhere' in the Contract Data. As long as it is clear.

What many people do is produce a table of buildings describing each one and explaining what is required in each, and what the constraints are for each (such as access). You can then put that table in the Service Information and the entry in the Contract Data can say something like 'The Affected Property is shown in Section 3 of the Service Information'. All of this of course will depend on exactly what maintenance or operations or facilities management you want the *Contractor* to carry out in these buildings.

231. Incomplete Contract Data (Contract Data)

We are a subcontractor under the ECS. The Contract Data (first bullet under general) states that retention and delay damages will apply. However, this has not been completed – how would this be interpreted?

The guidance warns that failure to complete fully will result in an incomplete contract – this is what you have here. There is no real solution here other than to agree to change the contract by agreement of the Parties (this would be a suitably authorised individual in your own company and within the main *Contractor*'s organisation) – this needs to be recorded in writing and signed by the two Parties (clause 12.3).

232. Do we need to name the *Project Manager*? (Contract Data)

I note in the ECC guidance notes on completing the Contract Data that individuals are named. Should this be the case or can we name a department? Also, what happens if the named individuals move on or need to delegate?

Yes, you should name the individuals – the ECC is all about having absolute clarity with communication and clear lines of authority. If those individuals are replaced then clause 14.4 requires that the *Employer* notifies the *Contractor* the name of the replacement.

Any delegation of duty by the *Supervisor* or *Project Manager* should be notified in accordance with clause 14.2. This should be clearly articulated – it is recommended that this is on a clause-by-clause basis.

Note that both these clauses use the term 'notify'. In accordance with clause 13.7 this means that the communication should be undertaken separately from other communications. Again, it is about clarity.

233. Works Information versus the *works* (Contract Data)

 I am competing the Contract Data for an ECC project – please can you advise on the difference between the *works* (second bullet) and the Works Information?

 The *works* should be a clear and succinct description of the *works*, e.g. the design and construction of a new oncology facility at the Newtown site.

The detail is contained in the Works Information. The Works Information is a defined term under the ECC (i.e. it has a contractual meaning). This is defined in clause 11.2(19) – it specifies and describes the *works*. It also states any constraints on how the *Contractor* Provides the Works.

234. Best practice in drafting the Contract Data (Contract Data)

We want to ensure that we comply with best practice in drafting the Contract Data as an *Employer* team. Can you suggest anything to ensure that we incorporate best practice and any relevant guidance?

The Works Information is one of the most important documents referenced from the Contract Data. The NEC published guidance on how to draft this in April 2013. It is recommended that the structure and drafting guidance is followed. This should ensure that the right terminology is used and this accords with ECC drafting principles.

Good practice in terms of the Contract Data would typically include striking through items (e.g. secondary Options and optional statements) that you do not intend to use. When circulating to a wider team this will highlight which options are not being used and may stimulate debate. Once a final version has been agreed these should be deleted in order to provide a 'clean' copy for the *Contractor*.

In addition, the optional statements should be 'cut' and 'pasted' into the relevant section as opposed to being left in their current position. For example, 'The Completion Date for the whole of the *works* is . . .' and 'The *Employer* is not willing to take over the *works* before the Completion Date' should both be included in the section titled 'time'.

235. The *period for reply* (Contract Data)

Please can you explain the *period for reply* – when is this applicable? As far as I am aware some of the clauses in the ECC have timescales – does the *period for reply* take precedence?

Clause 13.3 is perhaps the best starting point to explain the *period for reply*. Clause 13.3 states that, unless a timescale is stated in the contract, the *Supervisor*, *Contractor* and *Project Manager* all respond within the *period for reply*. Typically this is 2 weeks – sometimes this is less on projects that perhaps are fast track.

It is worth noting that differing *periods for reply* may be introduced, e.g. for design acceptance the *Project Manager* may give themselves more time, e.g. 3 weeks. They also may be contained in the Works Information. A note of caution – too long a period could make the project very difficult to deliver so a pragmatic approach is required.

236. Responding quicker than the *period for reply* (Contract Data)

What happens if the *Contractor* needs a decision quicker than the *period of reply*? Is the ECC *Project Manager* obliged only to respond within the *period for reply*?

The maximum time period for the *Project Manager* to respond (unless a timescale is defined in the contract) is the *period for reply*. In a spirit of mutual trust and co-operation this should be seen as the maximum.

If the *Contractor* requires a quicker turn around then the correct procedure in the contract is to notify an early warning. In a spirit of mutual trust and co-operation it is anticipated that the *Project Manager* would make all efforts to make a decision more quickly.

237. Contractual significance of the Risk Register (Contract Data)

Q Please can you explain the relevance of items that will be included in the Risk Register versus *Employer*'s risks in the ECC? What status do they have and how do they affect risk allocation?

A Items that will be included in the Risk Register should be listed in the Contract Data by both the *Contractor* and *Employer* at tender stage. They do not change the risk allocation. These are purely a proactive list of risks that, once the contract is signed, will be entered into the Risk Register and mitigation plans developed.

Items that are listed as *Employer*'s risks do change the risk allocation. If included and then they occur – these become a compensation event under clause 60.1(14).

The Risk Register is a defined term (clause 11.2(14)). It merely needs to include a description of the risk and mitigation plan. Many teams over-complicate this and lose track of practical mitigation measures.

So matters included in the Risk Register do not change risk allocation, additional *Employer*'s risks do.

238. The *defects date* versus the *defect correction period* (Contract Data)

Please can you explain the difference between the *defects date* and the *defect correction period*? Also, why are there potentially a number of *defect correction period*s in the ECC?

The *defects date* is a defined number of weeks after Completion of the whole of the *works*, e.g. 52 weeks (note, not after *sectional* Completion).

Effectively, this is the period within which the *Supervisor* would need to notify any Defects found and the *Contractor* is obliged to rectify within the *defect correction period*.

The *defect correction period* is therefore a defined period of time within which the Defect should be rectified. The *Contractor* cannot stockpile these and rectify in week 51 of 52 in order to gain retention release – they need to be dealt with on a case-by-case basis and within defined timescales.

239. Does the *defects date* change? (Contract Data)

We are using the ECC Option A contract on a develop-and-construct project. If an instruction for additional works is given after the Completion Date, and the instruction changes the Works Information and is a compensation event, what effect does this have on the *defects date*?

Assuming Completion of the whole of the *works* has been achieved, the instruction will have no effect on the *defects date*. This is because the *defects date* is defined as being 'xx' weeks after Completion of the whole of the *works* in the Contract Data part one. Of course 'xx' is the number of weeks inserted in the Contract Data in your particular contract. Completion is not affected as it has already been certified. The value of the compensation event will have to reflect the additional Defined Cost to the *Contractor* of carrying out and managing isolated works after its resources have left Site.

240. Optional statements – what are they? (Contract Data)

What are optional statements in Contract Data part one of the ECC and when should we consider these?

The optional statements are exactly that – you can decide to include them or not. These decisions should be in keeping with your contract strategy and cover 'what if' scenarios. For example, what if the *Contractor* achieves Completion prior to the Completion Date – are we willing to take over? If you are not, the optional statement should be included; if you are, then no need to include.

Other optional statements relate to payment timescales, additional *Employer*'s risks, Key Dates and insurances and, among others, the secondary Options.

These need to be carefully considered and should reflect project risks and key drivers/objectives.

241. Defining key people (Contract Data)

Q Do we have to let the *Contractor* define the key people or can we have a say on who they should be?

A Although Contract Data part two is completed by the *Contractor*, it can be altered by the *Employer*. You could, for example, define who the key people are, e.g. Site Agent, Planner, Commercial Manager, health and safety representative and the Design Manager.

You could also go further and define their qualifications and experience. This may be very important on some projects.

This has the advantage of defining what you see as the key attributes of a successful team on your project and also a level of consistency across the competing tenderers.

242. Modelling the Contract Data part two rates and percentages (Contract Data)

How should we evaluate the financial submission from the *Contractor* under Option C of the ECC? Should we just review the tendered total of the Prices or should we develop a cost model to assess other rates and percentages?

You should most certainly evaluate the rates and percentages in a cost model as you suggest. This provides sensitivity analysis and a rounded picture as to the most economical *Contractor*.

For example, *Contractor* A may submit a tendered total of the Prices at £1 million but his rates and percentages are treble those of *Contractor* B whose tender comes in at £1.01 million. If we estimate £300 000 of compensation events then *Contractor* B would be the most efficient. Appendix 4 of the ECC guidance notes provides a sample cost model – this would need developing for an Option C contract.

In order that all of the *Contractors* are aware of the cost model this would also need to be defined in the tender documents.

243. Optional statements (Contract Data)

Please can you explain the optional statement in Contract Data part two of the ECC – when would the *Contractor* provide Works Information for its design. Also, what happens if this is inconsistent with the *Employer*'s Works Information?

The *Contractor* may be required to provide Works Information for its design if tendering procedures require this.

If the Works Information prepared by the *Contractor* for its design is of a lower quality than the Works Information prepared by the *Employer*, then under clause 17.1, the *Contractor* or *Project Manager* are required to notify one another as soon as they become aware; the *Project Manager* will then decide the outcome. Assuming their instruction is for the *Contractor* to change its Works Information in order to comply with the *Employer*'s then this will not be a compensation event (clause 60.1(1), second bullet).

244. Completing the Contract Data (Contract Data)

We are using the ECC and are unsure about how the SCC and SSCC are implemented into the contract. For example, the data for the SCC and SSCC appear to be added at the end of the Contract Data part two by the *Contractor*.

My understanding is that these data relate to elements of the *works* outside the Working Areas (such as design and fabrication) and for say Equipment that is purchased. However, if our understanding is correct, we would also expect to see some words in the Contract Data part two along the lines of 'The SCC is in the document...' and 'The SSCC is in the document...'. There would then be a list of rates for various people and Equipment, etc. working on the Site; however, this does not appear to be the case. In short, we are unsure where the *Contractor* puts where, for example, a labourer costs £20 per hour or an excavator costs £35 per hour.

Both the SCC and the SSCC form part of the *conditions of contract*. They are incorporated into the contract by the wording of the first bullet point of Contract Data part one. What main Option you are on will determine which you use and what you use it for. The SCC (or SSCC in Options A and B) covers all Defined Cost and sets out what will be paid and how it will be paid. The information set out in Contract Data provides the information that is needed to make the SCC work, and is not just about work outside the Working Areas. People are paid for at what they cost, not on a quoted rate as you mention (see item 1 in the SCC and SSCC). Equipment is also paid at cost in the SCC (item 2), whereas in the SSCC it is paid for at the rates in the schedule referred to in the Contract Data part two with the percentage adjustment also set out there (item 2 of the SSCC).

245. Changes to the Contract Data part two percentages (Contract Data)

Under ECC Option C, is there any mechanism to implement a change to the Working Areas overheads percentage or to the percentage for people overheads to reflect a significant change in the *Contractor*'s resource profile on a project? In our case, there has been a significant increase in the use of direct labour to carry out work originally identified as being subcontracted.

The simple answer to your question is no. It is a figure that is included in the contract and which neither Party is able to change. Whether the *Employer* or the *Contractor* thinks it is too high or too low is irrelevant; you are both contracted to it. And if you are concerned about it we would ask you to give an honest answer to the following question – If the *Contractor* had come to you after the contract had been signed and said 'we are going to use a lot more Subcontractors than we thought so our Working Areas overheads percentage is too low and we want to increase it', what would your answer have been? Logic says that your answer (yes or no) should be the same as with your scenario.

In reality with Option C, this balance is part of the risks the Parties have agreed to share. You may pay the *Contractor* more in the way of the Price for Work Done to Date but the target does not increase, so you recover a proportion of that additional payment through the share mechanism (either more pain or less gain) at the end of the contract. How large that proportion is will, of course, depend on the share percentages and share ranges in the contract.

 **Some further thoughts and top tips**

- Ensure the Contract Data is completed in full. If incomplete we have an incomplete contract!
- Carefully select the main, secondary Options and optional statements and ensure they fit with the project risks and objectives.
- Develop a cost model that evaluates the Contract Data part two rates and percentages. Ensure this is clearly defined in the tender documents.
- Optional statements should be 'cut' and 'pasted' into the relevant sections (e.g. time/payment, etc).
- Ensure the Works Information (ECC), Service Information (TSC) or Scope (PSC) are appropriately drafted. Guidance provided by the NEC (April 2013) provides a clear structure and tips on drafting.